Flow Analysis of Computer Programs

THE COMPUTER SCIENCE LIBRARY

Programming Language Series
THOMAS E. CHEATHAM, *Editor*

Heindel and Roberto	LANG-PAK — An Interactive Language Design System
Wulf et al.	The Design of an Optimizing Compiler
Maurer	The Programmer's Introduction to SNOBOL
Cleaveland and Uzgalis	Grammars for Programming Languages
Hecht	Flow Analysis of Computer Programs

PROGRAMMING LANGUAGES SERIES

Flow Analysis of Computer Programs

Matthew S. Hecht

Department of Computer Science, University of Maryland

NORTH-HOLLAND·NEW YORK
NEW YORK • AMSTERDAM • OXFORD

Elsevier North-Holland, Inc.
52 Vanderbilt Avenue, New York, New York 10017

Elsevier Scientific Publishing Company
335 Jan Van Galenstraat, P.O. Box 211
Amsterdam, The Netherlands

© 1977 by Elsevier North-Holland, Inc.

Library of Congress Cataloging in Publication Data

Hecht, Matthew S
    Flow analysis of computer programs.
    (The Computer science library) (Programming languages series; 5)
    Bibliography: p.
    Includes index.
    1. Electronic digital computers—Programming. 2. Flowgraphs. 3. Algorithms. I. Title.
QA76.6.H4   001.6'1   77-881
ISBN 0-444-00210-3
ISBN 0-444-00216-2 pbk.

Manufactured in the United States of America

To Sara, my wife, Marci, my daughter, Joshua, my son,
and Jeffrey Ullman

CONTENTS

Preface xi

Acknowledgments xiii

Part I   Introduction and Preliminary Background 1

Chapter 1   Introduction 3

   1.1   Basic Concepts 3
   1.2   Motivation 5
   1.3   Overview 8
   1.4   Control Flow Analysis 9
       1.4.1   Quadruples: An Intermediate Program Representation 10
       1.4.2   The Call Graph 12
       1.4.3   The Control Flow Graph 12
   1.5   Data Flow Analysis 13
       1.5.1   Intraprocedural Data Flow Analysis: Example Problems 15
       1.5.2   Interprocedural Data Flow Analysis: Some Approaches 19
   1.6   Organization of This Book 22

Chapter 2   Preliminary Background 28

   2.1   Basic Mathematical Concepts and Notation 28
       2.1.1   Logic 28
       2.1.2   Proof Techniques 31
       2.1.3   Sets 32
       2.1.4   Relations, Functions 33
       2.1.5   Directed Graphs 34
   2.2   Comments about Algorithms 36
       2.2.1   Algorithms 36
       2.2.2   Data Structures 37
       2.2.3   Pidgin SIMPL 38

2.3	Introduction to Lattice Theory	40
	2.3.1 Posets, Chains	40
	2.3.2 Topological Sorting	41
	2.3.3 Lattices, Semilattices, Boolean Algebras	43
	2.3.4 A Fixed Point Theorem	46
	2.3.5 Properties of Operations on Semilattices	47

Part II Control Flow Analysis 51

Chapter 3 Flow Graphs, Dominance, Reducibility by Intervals, and Depth-First Spanning Trees 53

3.1	Flow Graphs	54
3.2	Dominance	55
3.3	Reducibility by Intervals	57
	3.3.1 Intervals	58
	3.3.2 Partitioning a Flow Graph into Intervals	60
	3.3.3 Reducibility	64
	3.3.4 Interval Order	65
3.4	Depth-First Spanning Trees	67

Chapter 4 Characterizations of Reducible Flow Graphs 72

4.1	Graph Transformations T1 and T2	73
	4.1.1 Collapsibility	73
	4.1.2 Equivalence of Reducibility and Collapsibility	77
4.2	A Forbidden Subgraph for Reducible Flow Graphs	78
4.3	Regions, Parses, and Backward Arcs	82
4.4	The DAG of a Reducible Flow Graph	91
4.5	Loop-Connectedness and Interval-Derived Sequence Length	95
4.6	Single-Entry Conditions	99
4.7	Subclasses of Reducible Flow Graphs	101

Chapter 5 Node Orders, Node Listings 105

5.1	Reasonable Node Orders	105
	5.1.1 Interval Order is Reasonable	106
	5.1.2 rPOSTORDER is Reasonable	107
5.2	A Dominator Algorithm for a Reducible Flow Graph	109
5.3	Node Listings	110
	5.3.1 Strong Node Listings	111
	5.3.2 Weak Node Listings	112

Contents

Chapter 6 Node Splitting — 114

6.1 One Simple Criterion — 114
6.2 A Definition of Node Splitting — 115
6.3 The Power of Splitting by Itself — 116
6.4 Why Node Splitting Is Not A Burning Issue — 120

Part III Data Flow Analysis — 125

Chapter 7 Some Simple Iterative Algorithms for Data Flow Analysis — 127

7.1 Representative, Basic Data Flow Analysis Problems — 128
   7.1.1 Available Expressions — 129
   7.1.2 Reaching Definitions — 132
   7.1.3 Live Variables — 132
   7.1.4 Very Busy Expressions — 132
   7.1.5 A Taxonomy — 133
7.2 Iterative Algorithm: Worklist Versions — 135
7.3 Iterative Algorithm: Round-Robin Version — 138
7.4 Iterative Algorithm: Node Listing Version — 143

Chapter 8 Interval Analysis — 146

8.1 Associating Certain Information with Arcs — 147
8.2 Notation — 147
8.3 An Efficient Data Structure for Interval Analysis — 149
8.4 Version 1: An Initial Description — 152
8.5 Version 2: A More Detailed Description — 153
8.6 Version 3: A Pigdin SIMPL Description — 155
8.7 Comments — 155

Chapter 9 Monotone Data Flow Analysis Frameworks — 160

9.1 Justification for a More General Setting — 161
9.2 Definition of a Monotone Framework — 163
9.3 The Meet Over All Paths (MOP) Solution — 169
9.4 Undecidability of the MOP Problem for Monotone Frameworks — 170
9.5 The General Iterative Algorithm — 173
9.6 A Dominator Algorithm for an Arbitrary Flow Graph — 179

Part IV A SIMPL Code Improver — 183

Chapter 10 A Modest Quad Improver for SIMPL-T — 185

- 10.1 Introduction — 186
  - 10.1.1 "Levels" — 186
  - 10.1.2 Design Goals — 187
- 10.2 SIMPL-T — 188
  - 10.2.1 Comments about SIMPL-T Compiler — 188
  - 10.2.2 A Synopsis of SIMPL-T Language Features — 188
  - 10.2.3 Quadruples of SIMPL-T — 191
- 10.3 Overview of the Quad Improver — 194
- 10.4 Interprocedural Analysis — 195
  - 10.4.1 The Call Graph — 195
  - 10.4.2 The Effect of a **CALL** Statement on the Calling Procedure — 196
  - 10.4.3 Summary of the Pre-pass — 198
  - 10.4.4 Transitive Closure — 198
  - 10.4.5 Aliasing — 201
- 10.5 Intraprocedural Analysis — 207
  - 10.5.1 The Equivalence Relation — 207
  - 10.5.2 Redundant Computations in Straight-Line Code — 208
  - 10.5.3 Redundant Computations in **WHILE** Loops — 210
  - 10.5.4 Redundant Computations in **IF** and **CASE** Statements — 211
  - 10.5.5 Invariant Code in **WHILE** Loops — 213
- 10.6 Discussion — 215

Bibliography — 219

Index — 225

PREFACE

This book presents a theoretical foundation for the pre-execution analysis of computer programs that is usually referred to as *control flow analysis* and *data flow analysis*. Flow analysis is a fundamental prerequisite for many important types of code improvement. In general, control flow analysis precedes data flow analysis. Control flow analysis is the encoding of pertinent, possible program control flow structure or flow of control, usually in the form of one or more graphs. Data flow analysis is the process of ascertaining and collecting information prior to program execution about the possible modification, preservation, and use of certain entities (such as values or various attributes of variables) in a computer program.

The primary goal of this book is to teach people algorithms to incorporate in code improvers. However, these algorithms do not perform various code improvements per se, but instead gather information prerequisite to many code improvements. Thus, the subject of this book is not code improvement, but only one constituent process used in many code improvers. The reader will be introduced to typical problems requiring flow analysis algorithms and the theoretical foundation for these algorithms.

The intended audience of this book includes: (a) practitioners who may want to implement some of these algorithms in existing or proposed compilers and interpreters; and (b) researchers, teachers, and students of programming languages who want to better understand the role of flow analysis, algorithms for data flow analysis, and the foundation of these algorithms. To aid the practitioners, we indicate at the beginning of each chapter (except Chapter 1, which contains the "big picture") those techniques and topics that are just theoretically interesting, and those that are useful in practice.

This book is organized into four parts, and ten chapters.

Part I supplies the introduction and preliminary mathematical background to flow analysis. Chapter 1 presents a general overview of flow analysis, including basic concepts, motivation, and example problems. The distinction between *intra*procedural and *inter*procedural flow analysis is explained. Chapter 2 contains most of the mathematical preliminaries for this study: definitions about directed graphs, and an introduction to semilattices. Pidgin SIMPL, the notation used for presenting algorithms, is also explained in Chapter 2.

Part II covers some topics in control flow analysis. Chapters 3 through 6 essentially contain a theory of reducible flow graphs. Chapter 3 defines the notion of reducibility in terms of intervals, Chapter 4 gives some characterizations of reducibilty, Chapter 5 introduces the concept of node listings, and Chapter 6 discusses node splitting. The ideas in Part II play important roles in the specification of algorithms for intraprocedural data flow analysis.

Part III presents some algorithms for intraprocedural data flow analysis, and a general framework in which to view intraprocedural data flow analysis problems. Versions of the practical iterative algorithm are covered in Chapters 7 and 9. Interval analysis is covered in Chapter 8. Frameworks are also introduced in Chapter 9.

Part IV introduces recursive descent flow analysis (sans flow graphs) for a simple, structured programming language. In Chapter 10 we discuss the design of a modest code improver for the SIMPL-T programming language.

This book is intended both for practitioners (implementors of programming language compilers and interpreters) and for researchers, teachers, and students of programming languages.

The practitioner will find Chapters 1, 7, 9, and 10 very useful, and maybe Chapters 3 and 8 (on interval analysis).

This book can be used as part of a graduate level course in computer science that explores various topics about programming languages. Such a course would be a sequel to the more established course on the theory of parsing, translation, and compiling. For example, a survey course can be divided into thirds, with one third on semantics, another third on correctness, and a final third on code improvement (or flow analysis). Another possibility is to rotate the subject material of a topics course with one entire semester on semantics, another semester on correctness, and another semester on code improvement. This latter approach has been used (to some extent) at the University of Maryland.

For a semester length course on code improvement, this book is useful but does not suffice. Additional material on the optimization of straight-line code is needed. Furthermore, a case study approach would be very valuable as part (not all) of such a course, as well as one or more limited implementation projects. Two ideas for projects in this course are: (1) a recursive descent code improver (which is limited in power) for a structured programming language; and (2) a pre-execution analyzer for APL that alleviates the run-time type-checking of APL where possible.

ACKNOWLEDGMENTS

In writing this book I have benefited from the helpful comments of many people. I want especially to thank Fran Allen, Susan Graham, John Kam, Barry Rosen, Ravi Sethi, Jeffrey Ullman, Len Vanek, and Marvin Zelkowitz. I have also received important comments and corrections from students who used these notes, among them Greg Frederickson, Roger Gifford, Leonie Penney, and Jeffrey Shaffer.

Special thanks go to Jeffrey Ullman who encouraged me throughout the preparation of this book, and to Jack Minker for his encouragement.

I wish to thank Laurie Richards, Eleanor Waters, and Alice Eichman for their excellant typing of the manuscript. In addition, I acknowledge the support provided by the Computer Center at the University of Maryland, and Kern Sibbald's document processor.

I am grateful to the University of Maryland for providing facilities for the preparation of this manuscript.

Part I
Introduction and Preliminary Background

Chapter 1
INTRODUCTION

This book presents a theoretical foundation for the pre-execution analysis of computer programs that is usually referred to as control flow analysis and data flow analysis. The primary goal of this endeavor is to teach people algorithms to incorporate in code improvers.

1.1 Basic Concepts

There is an important distinction between programming language translators and interpreters. "The general term *translator* denotes any language processor that accepts programs in some *source language* (which may be high- or low-level) as input and produces functionally equivalent programs in another *object language* (which may also be high- or low-level) as output. A *compiler* is a translator whose source language is a high-level language and whose object language is close to the machine language of an actual computer, either being an assembly language or some variety of machine language (relocatable or absolute)."[1] An *interpreter* is a language processor that executes its input source program itself, and does not produce an object program. One way to view the action of an interpreter is that it translates a source program statement prior to *each* time that statement is to be executed. FORTRAN, COBOL, ALGOL 60, PL/I, ALGOL 68, PASCAL, BLISS, and SIMPL-T are typically implemented using a compiler, whereas BASIC LISP, SNOBOL4, and APL are typically implemented using an interpreter.

For compiled language implementations, *run-time* (or *execution-time*) means during program execution, and *compile-time* (or *translation-time*) means prior to program execution and during program translation. The adjectives *static* and *dynamic* generally refer to compile-time and run-time respectively. Thus, for compiled language implementations, *pre-execution* means compile-time or static.

For interpreted language implementations, pre-execution means prior to interpretation. Pre-execution analysis of an interpreted language implementation is performed by one or more (beginning-to-end) passes over the source program prior to interpretation.

Because our subject is pre-execution analysis, we shall use adjectives such as 'possible' and 'potential' rather than 'actual' because we cannot be

[1] Pratt [1975], p. 21.

sure that an apparent control flow path will be realized by the program executing on some input.

The term *flow analysis* generally refers to compile-time analysis; that is, a compiled language implementation is assumed. However, we shall take it to mean pre-execution analysis. Flow analysis is usually dichotomized into control flow analysis and data flow analysis, where control flow analysis is, in general (although not necessarily), prerequisite to data flow analysis.

Control flow analysis is the encoding of pertinent, possible program "control flow structure" or "flow of control" for an ensuing data flow analysis. We envision a computer program as consisting of a collection of procedures. By a *procedure* we mean a compilable subprogram unit such as a main program, subroutine, or function. Execution of a computer program begins with the *main procedure* (AKA[2] *main program, driver, start proc*) that may call one or more other procedures or itself (each, one or more times), and each called procedure may do the same, and so forth. One representation of the calling relationships among the procedures of a program is a directed graph named a *call graph*, which shows "what may directly call what" one or more times.[3] The potential flow of control of each procedure is typically represented by a directed graph called a *control flow graph* or simply a *flow graph*, which is similar to a flow chart and depicts all possible execution paths.[4] The construction, representation, structure, and properties of such graphs are part of control flow analysis.

Data flow analysis is the pre-execution process of ascertaining and collecting information about the modification, preservation, and use of "quantities" in a computer program. Typically, (values of) variables are selected as the quantities under scrutiny, because they provide very fundamental information from which other information can be inferred. Not only are we interested in collecting information about variables holding values of type integer, real, character, or Boolean, but we are also interested in structure variables, procedure variables, label variables, type variables, and so forth, when they exist. Some information involving attributes of variables such as type, representation, range, or storage allocation can also be collected algorithmically.

Our real goal is *interprocedural data flow analysis* (i.e., program analysis). However, we can usually achieve this goal after suitable (but careful) preparation with a divide-and-conquer strategy (a procedure-by-procedure analysis) and restrict our attention to *intraprocedural data flow analysis*. For

---

[2]Here and henceforth we shall use 'AKA' as an acronym for 'also known as'.

[3]The problem of determining "what may directly call what" is not necessarily easy, due to procedure parameters and procedure variables.

[4]Again, the problem of determining a control flow graph is also not necessarily easy, due to label parameters, switches, and label variables.

Motivation

intraprocedural analysis, it is often convenient to post information gathered about a procedure at the nodes and arcs of the control flow graph of that procedure. One way to gather relevant information about a procedure is similar to the way gossip is propagated. Locally available information is initially posted, then subsequently propagated over the possible paths in the control graph. A "data flow analysis problem" should more appropriately be named an "information propagation problem".[5]

When control flow information is obvious, the construction of control flow graphs prior to data flow analysis is unnecessary. This situation occurs for languages such as BLISS and SIMPL-T, where a recursive descent approach suffices for many data flow analysis problems. When control flow information is somewhat difficult to determine, as in PL/I, other approaches are needed. For example, an initial conservative approximation to the control flow can allow data flow analysis to proceed, where the first goal of the data flow analysis is to obtain a more accurate representation of the control flow.

1.2 Motivation

There appears to be two general classes of applications of flow analysis. First, flow analysis can be used to derive information of use to human beings about a computer program. Such information aids in documenting, annotating, debugging, modifying, testing, and certifying programs. Second, flow analysis can be used to improve the efficiency of program execution. In addition to being the basis for most known code improvement algorithms, flow analysis can be used for certain "binding time" analyses to decrease execution time. We now sketch some of these applications.

(1) *Flow analysis can be used to derive information of use to human beings about a computer program.*

The programmer can be provided with a wealth of information about a program after flow analysis has been performed on it. For example, unreachable code, unused parameters, and variables that are used before being defined are identified by flow analysis. Annotations describing a program can be automatically inserted into its compilation listing. For example, local variables of FORTRAN subroutines whose values may depend on FORTRAN's local retention policy can be annotated. For each procedure definition and invocation, the parameters and other variables that are modified and used can be described, as well as other potential resulting

[5]Graham and Wegman [1976] suggest this.

invocations and transitive effects. For interactive debugging, all uses of each definition and all definitions affecting each use can be selectively extracted from a flow analysis data base. All of this information is especially helpful when using programs created by others.

Many program modifications are facilitated with flow analysis information. For example, changes in a procedure and its formal parameters may necessitate changes in each procedure referencing it or referenced by it. Such modification is easier when the transitive effects of a procedure are known.

Some automatic program restructuring may be possible too. For example, control flow simplification transformations may be desirable. As another example, if there is economy in eliminating procedure linkages in the intermediate representation of the program, some procedures can be expanded in-line and their calls eliminated, say for procedures that are called only once and act as macros. If such a procedure is needed some place else later, perhaps remodularization could be accomodated. Also, likely candidates for procedures could be suggested.

(2) *Flow analysis can be used to improve the efficiency of program execution.*

(2a) *Program Improvement.* Program improvement is the most important application of flow analysis. The goal here is to decrease the expected execution time of programs. Current approaches to removing useless code, eliminating redundant computations (such as common subexpressions), replacing run-time computations (such as constant propagation), performing code motions (such as removing invariant computations from loops), and providing register allocation information all employ data flow analysis as a preliminary step.

The additional time and cost of invoking a program improver is worthwhile for large repeatedly executed programs and very high-level language programs. For example, any systems program such as a compiler itself is repeatedly executed and, furthermore, subject to continual modification either for correcting errors or adding more user-oriented features. The availability of an automatic improvement mechanism frees a programmer of some otherwise tedious considerations both initially and after any program modification. For very high-level language programs, compile-time analysis and improvement inevitably translate (with a highly leveraged trade-off) into faster execution.

In addition, whatever the source of program blemishment, the analysis is the same. For example, a programmer may intentionally introduce redundancy at the source level to improve readability, or just by oversight.

Motivation

Furthermore, compilers usually generate redundant code when translating "heavy" referencing of multi-dimensional array elements.

(2b) *Binding Time Analysis.* The *binding* of a program object to a particular property, characteristic, or attribute is the choice of the property from a set of possible properties. *Binding time* is when such a choice is made; that is, when attributes or properties of an object are frozen.

Here are some examples. In LISP and SNOBOL4 we may dynamically create program text. Thus, for these languages the program itself is not in general frozen until run-time. In contrast, programs written in SIMPL-T are frozen at compile-time. If the types of all formal parameters in ALGOL 60 must be declared, then they are frozen at compile-time. If, however, the types of formal parameters in ALGOL 60 need not be specified and can vary during execution, then the binding time for the types of formal parameters in ALGOL 60 is at run-time. Some languages that allow the use of sets freeze the physical representation of sets when the language is defined or when the language processor is implemented. Other languages allowing the use of sets do not bind the physical representation until compile-time.

Postponement of binding times gives greater flexibility to a language user, but can adversely affect run-time efficiency. Pre-execution flow analysis can alleviate some of the implied run-time overhead. Without pre-execution flow analysis, certain bindings (e.g., run-time type checking in APL) are determined *in toto* at run-time. With pre-execution flow analysis, many such bindings can be determined prior to execution, greatly improving run-time efficiency.

We now mention some applications of pre-execution flow analysis to binding time problems.

Data flow analysis can be used to help determine types at compile-time for declaration-free languages. For example, this is the situation in SETL. In SETL, variables can be reused taking on different types in the same program, and an operator such as '+' may indicate either integer addition, set union, or tuple concatenation. Without knowledge of the types of objects in such languages prior to run-time, types must be dynamically determined at run-time, greatly slowing down execution. However, detailed information about the types of most objects can be determined by pre-execution analysis by the way such objects are defined and potentially used.

Pre-execution type-checking may be useful for APL. Bauer and Saal [1974] estimate that 80% of the rank, domain, and value checks required by a naive APL interpreter can be eliminated by pre-execution analysis.

Some automatic data structure selection is another application of pre-execution flow analysis. For example, consider a very high-level language in which a set definition and manipulation facility is provided. Rather than select a fixed representation for sets *a priori*, it is possible to postpone the representation decision until after a pre-execution analysis of the program has determined which operations are performed on different sets. Depending on the operations involved, a representation that is more appropriate than a default one can possibly be selected.

This same idea may be practical for data restructuring. For example, tables of data could be converted from a standard array representation to another efficient representation depending on the intended use and the translation benefit/cost ratio.

Jones and Muchnick [1976] have designed a pedagogical programming language called TEMPO to study binding time. In TEMPO, binding time is a parameter which may vary from one program to another, and even among the parts of a single program. They prescribe pre-execution flow analysis as the technique for determining whether or not certain bindings, although they appear to be at run-time, can be determined (or at least restricted) at compile-time. The following is a list of some problems they studied:

(a) Data types: Is run-time type checking needed and, if so, how can it be restricted?
(b) Storage management: Which variables require dynamic storage management and, if so, by what discipline (stack or heap)? What problems arise if array bounds need not be declared, any component of an array may itself be an array, and the number of dimensions of an array may vary?
(c) Name binding: Can the possible values of procedure variables or label variables be determined at compile-time?
(d) Parameter access methods: Can a more efficient method with the same behavior be substituted?
(e) Program text creation: When is the overhead for this aspect of a full language processor not necessary during execution?

Wegbreit [1974] considers other binding time problems in EL1, some of which can be analyzed by known data flow analysis techniques.

1.3 Overview

For the problems that we shall be considering, a simple textual scan of the source program (or an intermediate representation of it) is not in general sufficient to collect the desired information.

For the traditional approach to flow analysis (e.g., such as has been used with FORTRAN), an intermediate representation of a compiled language is subject to control flow analysis followed by data flow analysis. Information derived from data flow analysis is then used for some application (e.g., code improvement). (Thus, it is important to understand that data flow analysis is only a *preliminary* for an application such as code improvement.) For control flow analysis, a call graph is constructed, and a control flow graph for each procedure is constructed. For data flow analysis, an interprocedural analysis strategy is selected for handling **call** statements, and an algorithm for intraprocedural data flow analysis is selected to propagate information within a procedure. Most intraprocedural data flow analysis algorithms are based on posting and then propagating information about the arcs and nodes of a control flow graph.

The above scenario is not the only possible one for flow analysis; however, it is the one that we assume in most of this book (except for Chapter 10). In general, data flow analysis algorithms for programming languages with arbitrary jumps in control flow (caused by unrestricted **go to** statements) are based on control flow graphs. Sometimes when very restricted assumptions can be made about control flow, data flow analysis algorithms not requiring the construction of control flow graphs are possible. Such a situation is considered in Chapter 10. Also, flow analysis at the source level is sometimes very easy.

1.4 Control Flow Analysis

The purpose of control flow analysis is usually preliminary: to encode the flow of control of a program for use in the data flow analysis that follows.

Many compilers of high-level languages first translate a high-level source program into an intermediate-level program representation, then translate the intermediate-level program representation into a low-level program representation. For example, a high-level SIMPL-T program is translated into an intermediate form called "quadruples", which is then translated into machine language. Data flow analysis can be performed on any of the high-, intermediate-, or low-level representations of a program.

For certain applications (such as program verification), we would like to perform data flow analysis at the source level. If the flow of control is obvious at the source level, then no preliminary processing to glean flow of control information is necessary. (We shall explore one such situation in Chapter 10.)

For other applications (most notably code improvement), we typically perform data flow analysis at the intermediate level. If the flow of control is not obvious at the intermediate level, it must be determined.

Sometimes, program equivalence preserving flow of control transformations are applied to the control flow structure during control flow analysis to simplify the ensuing data flow analysis. For example, node splitting (discussed in Chapter 6), packaging new procedures, and expanding certain existing procedures in-line are some possibilities.

1.4.1 Quadruples: An Intermediate Program Representation

We assume that either (a) the flow of control of a program is immediate from the source level, perhaps after replacements of statements whose meanings involve control flow by more explicit statements; or (b) a program has been transformed by the "front end" of a compiler (e.g., the parser) into an intermediate form, such as *quadruples*, from which the flow of control is either explicit or very easily derived. For example, the parser of SIMPL-T[6] translates programs into quadruples.

Quadruples have four fields. The typical form of a quadruple is

$$(\underset{OP}{\langle \text{operator} \rangle}, \underset{A}{\langle \text{operand 1} \rangle}, \underset{B}{\langle \text{operand 2} \rangle}, \underset{R}{\langle \text{result} \rangle})$$

where we usually interpret it by saying that the $\langle \text{result} \rangle$ is obtained by applying the $\langle \text{operator} \rangle$ to the two arguments $\langle \text{operand 1} \rangle$ and $\langle \text{operand 2} \rangle$; that is, $R := OP(A,B)$. For unary operators, the second operand is empty. Temporary variables are often used for intermediate computations.

Example 1.1 The quadruples for $X := Y + Z$ are

$$(+, Y, Z, t1)$$

$$(:=, t1, , X)$$

where t1 is a temporary variable, and $:=$ is construed as a unary operator with the usual interpretation.

In addition to quadruples for arithmetic, logical, and relational operators, quadruples also exist for indexing arrays and calling subprograms.

Quadruples normally appear in the order in which they are to be executed. However, nonexecutable quadruples are used to supply necessary descriptive information, such as those delimiting a procedure or ending a parameter list. Flags indicating type and other information are usually associated with the operand and result fields.

Example 1.2 Figure 1.1 contains a procedure (actually a function) in Pidgin SIMPL[6] for Euclid's algorithm. Figure 1.2 contains a possible sequence of quadruples into which this function could be translated.

---

[6]SIMPL-T is described in Section 10.2, and Pidgin SIMPL is described in Section 2.2.3.

Control Flow Analysis

integer function GCD (**integer** P, **integer** Q)
```
/*-------------------------*
 *  Return the greatest common divisor of P and Q,  *
 *  where P and Q are positive integers.            *
 *-------------------------*/
```
    integer R
    $R := \text{REM}(P, Q)$ /* remainder of P divided by Q */
    while $R \neq 0$ **do**
        $P := Q$
        $Q := R$
        $R := \text{REM}(P, Q)$
    endwhile
return(Q)

Figure 1.1 A procedure for Euclid's algorithm.

OP-field	A-field	B-field	R-field
FUNC	GCD		
CALL	REM		t1
PARM	P	call-by-value	
PARM	Q	call-by-value	
ENDPARMS			
:=	t1		R
WHILE			
$\neq$	R	0 (constant)	t1
WHILETEST	t1		
:=	Q		P
:=	R		Q
CALL	REM		t1
PARM	P	call-by-value	
PARM	Q	call-by-value	
ENDPARMS			
:=	t1		R
ENDWHILE			
RETURN	Q		
ENDFUNC			

Figure 1.2 A possible sequence of quadruples for Figure 1.1.

1.4.2 The Call Graph

Most computer programs consist of a collection of procedures. Typically, execution begins with the start proc, which in turn may reference (or call, invoke) one or more other procedures or itself (each, one or more times), and each referenced procedure may do the same, and so on. An *internal procedure* is one that is present and available for analysis, whereas an *external procedure* is referenced but missing. For example, the so-called *intrinsic* (or *built-in*) functions of most programming languages, such as input/output routines, can be considered external procedures although complete information about each is known. Other external procedures, about which complete information is not known, may be referenced. (The desirability of "separate compilations" and external procedures is generally accepted.)

One representation of the referencing relationships among the procedures of a program is a directed graph named a *call graph*. Each node of a call graph corresponds to exactly one procedure of a program, including the start proc, and each arc (x,y) represents one or more references in the procedure represented by node x of the procedure represented by node y. Note that a call graph represents procedure references only, and not returns from procedures.

To construct a call graph, we assume that the start proc and all procedures transitively referenced by it are easy to identify. (Referencing information about external procedures is stored in an accessible "mail box" or data base.) For example, suppose we have a Pidgin SIMPL program with no external procedures (except intrinsics) that has been completely translated into a sequence of quadruples. The OP-field of the last quadruple is START, and the A-field is the name of the start proc. The A-field of any quadruple whose OP-field is CALL is the called procedure, and the (suitably delimited) procedure containing this call is the calling procedure.

Example 1.3 Suppose the GCD procedure in Figure 1.2 is to be tested. The start proc of the testing program is a procedure named TESTER. TESTER references intrinsic procedures READ and WRITE, as well as GCD. GCD references REM. Figure 1.3 presents the call graph of this testing program, which happens to be a tree.

1.4.3 The Control Flow Graph

From each intermediate form procedure or directly from each procedure itself we partition its "statements" into maximal groups such that no transfer occurs into a group except to the first statement in that group, and once the first statement is executed, all statements in the group are

Data Flow Analysis

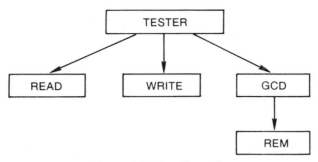

Figure 1.3 A call graph.

executed sequentially. (Each **call** statement or reference to a procedure is treated like an assignment statement for this partitioning.) Such a group of statements is called a *basic block*, or simply a *block*. (Do not confuse this block with a **begin**...**end** block of, say, ALGOL 60.)

From the basic blocks of a procedure we construct a directed graph that resembles a flow chart of the procedure. Each node of this graph corresponds to a basic block of the procedure, and there is an arc (x,y) from block x to block y if control can potentially transfer from block x to block y at run-time. This graph is called a *control flow graph*, or simply a *flow graph*. The node corresponding to the block containing the first statement of a procedure is called the *initial node*. Any node whose corresponding block contains a **return** statement is called a *terminal node* or *exit node*.

Any node of a control flow graph that is not accessible from the initial node (or any procedure of a call graph that is not transitively referenced by the start proc) represents inaccessible code, and can be removed from the program.

In addition, there is no loss in generality in assuming that a control flow graph has exactly one initial node with no entering arcs. If the initial node has one or more entering arcs, just create a new initial node with one arc leaving it and entering the old initial node. For flow graphs that have more than one "entry" node, add a new initial node and an arc from this new node to each entry node.

Example 1.4 A flow chart and control flow graph of Euclid's algorithm are shown in Figure 1.4.

1.5 Data Flow Analysis

Data flow analysis can be described as the pre-execution process of ascertaining and collecting information about the possible run-time modification, preservation, and usage of certain quantities in a computer pro-

Introduction

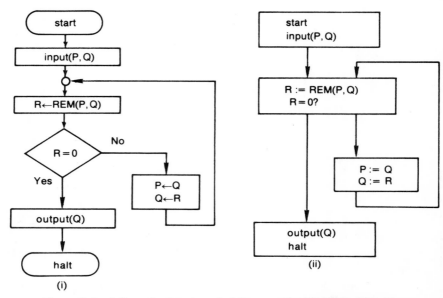

**Figure 1.4 A flow chart and control flow graph of Euclid's algorithm.
(i) Flow chart; (ii) control flow graph.**

gram. Usually "values of variables" are selected as the quantities under scrutiny, because they provide very fundamental information.

We can distinguish among several "levels" of analysis:

(a) statement, or quadruple;
(b) basic block (intrablock);
(c) procedure (intraprocedural); and
(d) program (interprocedural).

Some authors describe statement and intrablock analysis by the term *local*; intraprocedural and interprocedural analysis by the term *global*. Local flow analysis, and the flow analysis of programs without loops (or recursion) is easy and has been done elsewhere.[7] In this book, we usually mean 'intraprocedural' when we use the phrase 'data flow analysis'.

Data flow information is collected hierarchically. If we know for each statement what variables have their values modified, preserved, and used if control flow proceeds through the statement, then we can infer corresponding information for each block in a control flow graph of a procedure.

[7] For example, see Aho and Ullman [1973] and Cocke and Schwartz [1970].

Data Flow Analysis

Similar information about procedures leads to information about the program as a whole. However, the collection of the intraprocedural information is somewhat complicated by control flow. Even information about particular individual statements, such as **call** statements, may not be immediately available.

Incidentally, techniques generally called global data flow analysis can be employed to determine certain control flow information that may not be immediately available as assumed. For example, determining the potential values or sets of values of label variables and procedure variables, if such variables exist, may be possible.

1.5.1 Intraprocedural Data Flow Analysis: Example Problems

We introduce intraprocedural data flow analysis by describing two typical problems and indicating some of the applications of the information derived. These two problems are called "reaching definitions" and "live variables". From the solutions of these two problems, the solutions to two other problems, called "live definitions" and "definition-use chaining", are easily obtained.

Because we intend the following as examples, and not statements of the problems in their most general forms, we select variables and definitions of variables as the quantities under consideration rather than something more abstract.

We assume that (a) all relevant local data flow information is available for a particular procedure; (b) any variable "aliasing"[8] is known; and (c) we have a control flow graph for this procedure. We denote the control flow graph by the triple $G=(N,A,s)$, where (N,A) is a directed graph, $s \in N$ is the initial node with in-degree 0, and there is a path from s to every node.

Each problem below is described in terms of this control flow graph. We shall associate some information with the top of each node (pictorially following the "join" of the entry arcs) and other information with the bottom (pictorially preceding the "fork" of the exit arcs). The suffixes -TOP and -BOT distinguish these. The sets of information defined can be conveniently represented by bit vectors.

---

[8]Two distinct variables are *aliases* when they refer to the same memory location. Thus, whenever one variable is modified, so is the other. For example, the EQUIVALENCE statement of FORTRAN introduces static aliasing. Dynamic aliasing is possible too. For example, when procedure P with call-by-reference formal X is called with global actual A, then X and A are aliases in P at run-time. Also, if procedure P has two call-by-reference formals X and Y, and P is called with global actuals B and B (identical actuals), then X, Y, and B are aliases in P at run-time.

(a) *Reaching definitions*. For this problem we would like to determine the sets RDTOP(x) and RDBOT(x) of variable definitions that can "reach" the top and bottom of each node x in the control flow graph G. Actually, only RDTOP(x) is useful.

By a variable *definition* we mean a statement that can (potentially) modify the value of a variable, such as an assignment statement or a **read** statement. A path in a control flow graph is called *definition-clear* with respect to a variable v iff there is no definition of v on that path. A definition d of a variable v in a node x is said to *reach* the top (respectively: bottom) of node y iff d occurs in node x and there is a definition-clear path for v from d to the top (respectively: bottom) of node y.

Two sets of local information are known at the bottom of each node x; namely, XDEFS(x) and PRESERVED(x).

A *locally exposed definition* or a *locally generated definition* is the last definition of a variable in a node. For each node x, let XDEFS(x) be the set of locally exposed definitions of node x. Intuitively, these are the definitions "created" or "generated" by each node as far as the outside world is concerned.

A definition d of a variable v is said to *kill* all definitions of the same variable that reach d. Equivalently, any definition of a variable v that reaches the top of a node x, and there is no definition of v in x, is said to be *preserved* by x. For each node x, let PRESERVED(x) be the set of definitions preserved by node x. Intuitively, these are definitions "transmitted" by each node.

There is one more item to clear up before we complete the description of the reaching definitions problem. Let OUTSIDE be the set of definitions that reach s from outside the procedure. (Remember, this is possible if the procedure under analysis is called in an environment in which definitions do reach the **call** statement.) To simplify the exposition here, we assume (WLOG)[9] that OUTSIDE = $\varnothing$. To justify this assumption, if OUTSIDE $\neq \varnothing$ then we create a new control flow graph $G' = (N \cup \{s'\}, A \cup \{(s',s)\}, s')$, where $s' \notin N$, RDTOP(s') = $\varnothing$, XDEFS(s') = OUTSIDE, and PRESERVED(s'), = $\varnothing$.

RDTOP and RDBOT can be succinctly described by the equations in Figure 1.5. Note that each of RDTOP and RDBOT can be described (and solved) without the existence of the other. RD4 is RD2 substituted into RD1, and RD6 is RD1 substituted into RD2. (See Chapter 7 for RD3.)

(b) *Live variables*. For this problem we would like to determine the sets LVTOP(x) and LVBOT(x) of variables that are "live" or may be used after control passes the top or bottom of node x in the control flow graph

[9]'WLOG' is an acronym for 'without loss of generality'.

Data Flow Analysis

Equations for RDTOP and RDBOT:
RD1. For each node x,

$$\text{RDTOP}(x) = \bigcup_{y \in \text{PRED}(x)} \text{RDBOT}(y).$$

(If $\text{PRED}(s) = \emptyset$, then $\text{RDTOP}(s) = \emptyset$.)

RD2. For each node x,

$$\text{RDBOT}(x) = [\text{RDTOP}(x) \cap \text{PRESERVED}(x)] \cup \text{XDEFS}(x).$$

Equations for RDTOP alone:
RD4. For each node x,

$$\text{RDTOP}(x) = \bigcup_{y \in \text{PRED}(x)} [(\text{RDTOP}(y) \cap \text{PRESERVED}(y)) \cup \text{XDEFS}(y)].$$

(If $\text{PRED}(s) = \emptyset$, then $\text{RDTOP}(s) = \emptyset$.)

Equations for RDBOT alone:
RD5. $\text{RDBOT}(s) = \text{XDEFS}(s)$
RD6. For each node $x, x \neq s$,

$$\text{RDBOT}(x) = \left[\left[\bigcup_{y \in \text{PRED}(x)} \text{RDBOT}(y)\right] \cap \text{PRESERVED}(x)\right] \cup \text{XDEFS}(x).$$

Figure 1.5 Equations for "reaching definitions".

G. Both $\text{LVTOP}(x)$ and $\text{LVBOT}(x)$ are useful. For example, they enable us to delete useless STOREs.

A variable is said to be *used* iff it is referenced without modification, as by a **write** statement or when it is an operand of a computation. A variable v in node y is said to be *live* or *alive* (antonym: *dead*) at the bottom (respectively: top) of node x iff a use of v occurs in y and there is a definition-clear path for v from this use backward to the bottom (respectively: top) of node x. That is, v is alive at the bottom of x iff there is a definition-clear path for v from the bottom of x to a use of v.

Again, two sets of local information are known at each node x; namely, $\text{XUSES}(x)$ and $\text{NOTDEFINED}(x)$.

A *locally exposed use* of a variable v is a use of v in a node x, which is not preceded by a definition of v in x. For each node x, let $\text{XUSES}(x)$ be the set of variables with locally exposed uses in node x. Intuitively, if we look into the top of a node, each of these uses would produce a run-time error if no definition for that use reached the node.

For each node x, let NOTDEFINED(x) be the set of variables that are not defined in node x.

For the "live definitions" problem we need to clarify two points. Recall that an exit node of G has out-degree 0. Also, uses from outside the procedure under analysis may be live at the bottom of each exit node. Analogous to the previous problem, to simplify the exposition here we assume (WLOG) that for each exit node w of G, LVBOT(w) = $\emptyset$—for otherwise we could modify G by adding new appropriate exit nodes.

For the second point, we assume that each variable is encoded as the subset of definitions that define that variable. Then NOTDEFINED(x) is just PRESERVED(x). Also, sets from both the "reaching definitions" and "live variables" problems can now be meaningfully intersected.

LVTOP and LVBOT can be defined by the equations in Figure 1.6. Note that RD4 for RDTOP is similar to LV4 for LVBOT. Also, the reaching definitions problem is a "top-down" problem, whereas, the live variables problem is a "bottom-up" problem. (See Chapter 7 for LV3.)

Equations for LVTOP and LVBOT:
 LV1. For each node x,

$$\text{LVBOT}(x) = \bigcup_{y \in \text{SUC}(x)} \text{LVTOP}(y).$$

(If SUC(w) = $\emptyset$ for any node w, then LVBOT(w) = $\emptyset$.)
 LV2. For each node x,

$$\text{LVTOP}(x) = [\text{LVBOT}(x) \cap \text{PRESERVED}(x)] \cup \text{XUSES}(x).$$

Equations for LVBOT alone:
 LV4. For each node x,

$$\text{LVBOT}(x) = \bigcup_{y \in \text{SUC}(x)} [(\text{LVBOT}(y) \cap \text{PRESERVED}(y)) \cup \text{XUSES}(y)].$$

(If SUC(w) = $\emptyset$ for any node w, then LVBOT(w) = $\emptyset$.)

Equations for LVTOP alone:
 LV5. For each exit node w, LVTOP(w) = XUSES(w).
 LV6. For each nonexit node x,

$$\text{LVTOP}(x) = \left[\left[\bigcup_{y \in \text{SUC}(x)} \text{LVTOP}(y)\right] \cap \text{PRESERVED}(x)\right] \cup \text{XUSES}(x).$$

Figure 1.6 Equations for "live variables".

Data Flow Analysis

(c) *Live definitions*. A definition d is *live* or *active* at the top of node x iff $d \in \text{RDTOP}(x) \cap \text{LVTOP}(x)$.[10] That is, not only does the definition reach the top of x, but it can potentially be used. Note that the "live definitions" problem is different from the "live variables" problem and that the former gives "stronger" information.

The sets of live definitions are useful when assigning index registers: registers holding dead definitions can be reused immediately.

(d) *Definition-use chaining*. Suppose we know $\text{RDTOP}(x)$, $\text{LVBOT}(x)$, $\text{XDEFS}(x)$, and $\text{XUSES}(x)$ for each node x. By considering both $\text{RDTOP}(x)$ and $\text{XUSES}(x)$ together, we can establish a pointer from each use in $\text{XUSES}(x)$ to the location of zero or more definitions in $\text{RDTOP}(x)$. A similar association can be established between $\text{LVBOT}(x)$ and $\text{XDEFS}(x)$. This double linking is called definition-use chaining. Combined with other local information, we know, for a given definition, what uses might be affected by it and, for each use, what definitions can affect it.

Such information is useful for dead code elimination, constant propagation, and error detection. For example, if a given definition affects no uses, that definition can be eliminated. If all definitions reaching a particular use are the same constant, we can use this fact to perform constant propagation. (Again, the important point is that flow analysis collects information useful for or prerequisite to code improvement. It does not accomplish the complete improvement.)

1.5.2 Interprocedural Data Flow Analysis: Some Approaches

Intraprocedural data flow analysis presupposes that local information is immediately available, and *a fortiori*, easy to determine. Unfortunately, this is usually not the case, especially for **call** statements. In this section we briefly introduce several possible approaches to analyzing a program consisting of a collection of procedures.

One such method is termed the *worst case method*. According to this method we make "worst case" assumptions about the effects of **call** statements, parameter aliasing, and the external environment of each procedure. Despite the myopic view taken by this approach, we must resort to such a strategy, for example, for external procedures, when missing necessary information.

Another interprocedural analysis strategy is termed the *complete expansion method*. For this method we first expand each procedure in-line into

---

[10]It is assumed here and in the next problem to be described that definitions and uses are suitably encoded so that they can be meaningfully intersected.

20   **Introduction**

the start proc, resolving local variable name conflicts, updating local variable declarations, and replacing formal parameters by the actual arguments specified in each **call** statement. Then we just analyze the remaining one procedure by some intraprocedural method, since it contains no **call** statements. Although this method may lack beauty, it is a simple idea and sometimes fruitful. Should the call graph of the program have any cycles, however, the expansion phase would never terminate. Also, external procedures are not available.

If any procedure in the call graph is recursive, the call graph will contain a cycle. But it is also possible for a call graph to contain a cycle even when no procedure is recursive.

Example 1.5 Consider the call graph in Figure 1.7(i). Suppose neither B nor C is recursive, and that C only calls B if A calls C. A more appropriate representation of this particular situation is given by the node-labeled call graph in Figure 1.7(ii).

If the call graph is a DAG,[11] then complete in-line expansion is possible. If a procedure is called more than once, it will be analyzed repeatedly in varying contexts. But still, we face the possibility of an explosion in the size of the new start proc if procedures that were previously heavily called were large. Also, after complete expansion, we may have lost any advantage of a divide-and-conquer strategy for data flow analysis already established by the foregone modularization. That is, repeated analysis of each procedure called more than once in each unique context can be expensive.

Another interprocedural analysis strategy is termed the *one-pass method*. Each procedure is analyzed exactly once. Extremely conservative (but enlightened by preprocessing) "worst case" assumptions are made to

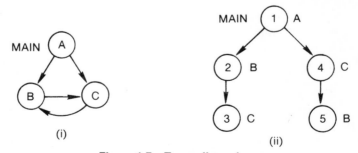

Figure 1.7 Two call graphs.

[11]'DAG' is an acronym for 'directed acyclic graph'. Definitions for the terms DAG, topsort, acyclic condensation, and SCC (strongly connected component) can be found in Chapter 2.

Data Flow Analysis

supply the needed but missing information, since one analysis must apply to all calling contexts. This method is rapid, but can give very weak information.

The order in which procedures are analyzed in the one-pass method determines the availability of information. First suppose that the call graph is a DAG. If we analyze the procedures according to a topsort[11] of the call graph, then we have to make worst case assumptions about **call** statements. What we would like to know is the transitive effects of all the procedures that a given procedure x calls before we analyze procedure x. Thus the more appropriate procedure analysis order for the one-pass method is a reverse topsort of the call graph. Using this order, each procedure is analyzed only after all procedures that it calls have been analyzed. Now local information due to **call** statements is known. However, what assumptions do we make about the outside information "reaching" the top or bottom of each procedure? For example, suppose our program consists of two procedures, A and B, where A calls B and B calls no procedures. If we analyze each procedure exactly once, and B first, we must make "worst case" assumptions when we analyze B, perhaps enlightened by preprocessing, about what reaches B when it is called in A.

If the call graph has cycles, then the acyclic condensation[11] of the call graph can be formed. Since the acyclic condensation of the call graph is a DAG, its SCCs[11] can be analyzed in a reverse topsort order. Procedures in each SCC can be analyzed in any order. All information obtained from analyzing the procedures of an SCC must be pooled together. If procedure x calls procedure y, then the information in the pool of the SCC containing procedure y is used. We are assuming some preprocessing here.

One class of interprocedural analysis strategies is termed the *multi-pass methods*. Begin with an initial estimate about unknown information and make a pass through the procedures in some appropriate order, producing a revised estimate. Then using the revised estimate, iterate again. Iteration continues until the information "stabilizes".

Another strategy is called *symbolic flow analysis*. Instead of answering a question about data flow through a procedure with one bit of information, we answer it with a Boolean expression that uses symbolic variables for the data flow questions associated with calls within the body. Thus, rather than reanalyzing procedure bodies once for each call as in the multi-pass methods, we just evaluate the expressions for each call.

Finally, there are other methods, most of which are composite, which we classify as *mixed-strategy methods*. For example, the worst case method can be used for external procedures. A partial expansion can be performed by fully expanding only procedures of very short length or that are called exactly once. After this, a multi-pass method or symbolic flow analysis can be used. Some recursive procedures can even be rewritten as iterative nonrecursive procedures in part of an "expansion" phase.

There is a growing body of papers on interprocedural data flow analysis. Our cursory survey just scratches the surface of this important area.

1.6 Organization of This Book

The ten chapters of this book are divided into four parts.

Part I is titled Introduction and Preliminary Background. It contains the first two chapters. Chapter 1 introduces the reader to control flow analysis and data flow analysis. Representative intraprocedural data flow analysis problems are presented, and some approaches to interprocedural data flow analysis are discussed here (and also in Chapter 10). Chapter 2 covers the basic mathematical concepts and notation employed in this study. Pidgin SIMPL, our notation for presenting algorithms, is described in this chapter and some introductory lattice theory is included.

Part II is titled Control Flow Analysis. It contains Chapters 3 through 6. In these chapters we present a theory of "reducible" flow graphs that serves as a foundation for many intraprocedural data flow analysis algorithms. Chapter 3 introduces the following concepts: flow graphs, dominance, reducibility by intervals, and depth-first spanning trees. Chapter 4 contains some characterizations of reducible flow graphs. Chapter 5 introduces node orders and node listings. Chapter 6 discusses node splitting.

Part III is titled Data Flow Analysis. It contains Chapters 7 through 9. Chapter 7 presents variations of the iterative algorithm for solving intraprocedural data flow analysis problems such as "available expressions" and "live variables". Chapter 8 details the interval analysis approach for solving intraprocedural data flow analysis problems. Chapter 9 introduces the notion of a monotone framework, which elegantly models general intraprocedural data flow analysis problems.

Part IV is titled A SIMPL Code Improver and contains Chapter 10. In it we sketch some ideas for the design of a code improver for SIMPL-T. The purpose of the chapter is to explain how some data flow analysis problems can be solved by a slightly modified recursive descent approach (without control flow graphs), when the underlying programming language has very restricted control flow.

How does everything fit together? Recall from the overview in Section 1.3 that we must select an interprocedural strategy (essentially to handle **call** statements), and an intraprocedural algorithm to process individual procedures. The control flow analysis performed is very dependent on the intraprocedural data flow analysis algorithm selected.

Organization of This Book

Some interprocedural data flow analysis strategies are discussed in Sections 1.5.2 and 10.4. Here is a list of these strategies.

(a) Worst case method.
(b) Complete expansion method
(c) One-pass method.
(d) Multi-pass method.
(e) Symbolic flow analysis.
(f) Mixed-strategy methods.

Here is a list of some known intraprocedural data flow analysis algorithms.

(a) Strongly connected region approach of Allen [1969].
(b) Interval analysis of Allen and Cocke [1976].
(c) Iterative algorithm of Kildall [1972] and Ullman [1973].
(d) T1-and-T2-based algorithm of Ullman [1973].
(e) Node listing algorithm of Kennedy [1975a].
(f) T3-T4-and-T5-based algorithm of Graham and Wegman [1976].
(g) Graph grammar approach of Farrow, Kennedy, and Zucconi [1976].
(h) Attribute grammar approach of Jazayeri [1975].
(i) Slightly modified recursive descent approach of Wulf *et al.* [1975], Zelkowitz and Bail [1974], and Hecht and Shaffer [1976].
(j) Symbolic flow analysis of Rosen [1976].
(k) Elimination methods (see Backhouse [1976], for example).

In this book we present approaches (b), (c), (e), and (i) from the above. We discuss interval analysis in Chapter 8. Two topics are prerequisite to an understanding of interval analysis: intervals in Chapter 3, and node splitting in Chapter 6. The iterative algorithm is discussed in Chapters 7 and 9. The material on node orders in Chapter 5 is relevant to this algorithm. The node listing algorithm is also presented in Chapter 7. Node listings are discussed in Chapter 5. A slightly modified recursive descent approach is presented in Chapter 10. Chapter 4 contains material fundamental to the understanding of (d), (f), and (g) above. We shall not discuss elimination methods in this book.

Which of the above eleven algorithms are most important? Why are only four of the above eleven algorithms presented in this book? Which algorithm should you select for your application?

First let us talk about relative importance and selection of algorithms for presentation. The iterative algorithm seems to be the most important one, because it is applicable to an extremely large class of data flow analysis problems and because it is uninfluenced by the reducibility of the underlying flow graph. Because the node listing approach is a simple modification of the iterative algorithm, we have included it for presentation along with the iterative algorithm. Interval analysis supersedes the strongly connected region approach. We have included interval analysis in our presentation for several reasons: it is very representative of algorithms in this area that are based on reducing a flow graph by simple transformations; it is incorporated in an existing experimental code improver for PL/I; and other algorithms for data flow analysis presume an understanding of interval analysis, usually as a comparison (i.e., it is a benchmark). The slightly modified recursive descent approach is important because of its simplicity. It was included for this reason and because it is a good contrast to the methods based on flow graphs.

The decision not to include the other algorithms was partly arbitrary (to limit the length of this book) and partly due to the fact that some of these approaches are new. Ullman's algorithm and the Graham-Wegman algorithm are each asymptotically faster than interval analysis. The approaches based on graph grammars, attributed grammars, and symbolic flow analysis have been developed very recently.

The decision as to which algorithm to select for a particular application is based on at least three things: the difficulty of the data flow analysis problem, the features of the programming language, and the other problems requiring attention. Not all data flow analysis problems are the same. Some are more difficult than others, and may require a more powerful algorithm such as the general iterative algorithm. Certain features of the programming language under scrutiny may rule out some approaches. For example, for programming languages with arbitrary **go to** statements such as FORTRAN, PL/I, and SETL, we cannot generally use the recursive descent approach. The third point is that some approaches provide information that is necessary to solve other problems. For example, interval analysis identifies the "loops" of a program, and this is also important for moving code. Loops are not explicitly identified by the iterative algorithm.

As is the case with sorting algorithms, no one intraprocedural data flow analysis algorithm is best in any absolute sense. The practitioner must fit the algorithm to the application. However, we strongly recommend the general iterative algorithm, based on round-robin node visitation (see Chapters 7 and 9). In the following we shall try to explain the advantages and disadvantages of each intraprocedural data flow analysis algorithm that we present. Also we shall try to point out which techniques and topics are just theoretically interesting, and which are useful in practice.

Exercises

1.1 When referring to the value of a variable during pre-execution data flow analysis, which of the following phases are equivalent, and which are negations of others?
   (a) cannot be modified   (f) must be preserved
   (b) may be modified   (g) cannot be used
   (c) must be modified   (h) may be used
   (d) cannot be preserved   (i) must be used
   (e) may be preserved

1.2 What features of (LISP and SNOBOL4 thwart pre-execution analysis? Can data flow analysis be used to determine when such features cannot occur in a particular program written in one of these languages?

1.3 Suppose we want to determine the set of variables whose values may be modified by an assignment statement in PL/I. How may the presence of ON-units in a PL/I program complicate this task? Should ON-units defined in external modules (separate compilations) be considered?

1.4 Design a program that will scan the quadruples of a SIMPL-T program (see Sections 1.4.1 and 10.2) and construct a call graph.

1.5 Consider a programming language such as PASCAL or ALGOL 60 that has procedure parameters. Let P be a procedure with a procedure parameter X. Design an algorithm to determine at compile-time the set of possible actual procedures that X may be associated with at run-time. Note that if procedure Q has procedure parameter Y, then in Q there may be a call of P(Y). Does the scope rule of block-structure affect the solution?

1.6 Let MAYCALL be a relation on procedure names of a PASCAL program, where P MAYCALL Q means that P may call Q at run-time. Using your analysis from Exercise 1.5, design a program to compute the MAYCALL relation for a PASCAL program.

Bibliographic Notes

Pratt [1975] is a good survey of programming languages and run-time environment. For example, it covers the notions of compiler versus interpreters and compile-time versus run-time.

In 1961 Vyssotsky [1973] implemented a compile-time diagnosis facility using (intraprocedural) control flow analysis and data flow analysis as a postprocessor of a Bell Laboratories 7090 FORTRAN II compiler. After constructing a control flow graph (for each procedure) of the source program, Vyssotsky (evidently) solved the "reaching definitions" problem, using bit vectors to represent sets. By combining this global (intraprocedural) information about definitions reaching each block with the local (intrablock) exposed uses, potential errors were inferred, thus generating appropriate messages. (In our notation, each use in $XUSES(x) - RDTOP(x)$ was annotated, for each node x.) "Worst case" assumptions were made for subprogram invocations.

Allen [1969] and Lowry and Medlock [1969] are two excellent papers describing the use of intraprocedural flow analysis with flow graphs for code improvement of FORTRAN programs. Good reference books discussing code improvement using flow analysis are Cocke and Schwartz [1970], Aho and Ullman [1973], Schaefer [1973], and Aho and Ullman [1977]. Allen [1975a] contains a bibliography (up to December, 1975) program optimization. (We shall reference many more papers and isolate specific contributions in later chapters.)

For general motivation of flow analysis, see Allen [1974b], Osterweil and Fosdick [1974], Fosdick and Osterweil [1976], and Jones and Muchnick [1976]. Binding time analysis is discussed in Wegbreit [1974] and Jones and Muchnich [1976], for example. Pre-execution analysis for type determination is discussed in Bauer and Saal [1974], Tenenbaum [1974], and Fong, Kam, and Ullman [1975]. Automatic data structure selection using data flow analysis can be found in Schwartz [1975].

Intermediate forms of computer programs (such as quadruples) are discussed in Gries [1971], and Aho and Ullman [1977], for example.

The notion of a call graph has been around for a while. Some references using this notion are Strong, Maggiolo-Schettini, and Rosen [1975] and Allen [1974a].

Bibliographic Notes

Cocke and Schwartz [1970] and Aho and Ullman [1973] cover methods for local (intrablock) data flow analysis and data flow analysis of programs without loops. These methods are not included in this book.

The equations for "reaching definitions", "live variables", and "live definitions" occur in a large number of papers. Two recent and lucid presentations are given in Cocke and Allen [1975], and Aho and Ullman [1977].

Some excellent references on interprocedural data flow analysis are Spillman [1971], Allen [1974, 1975b], Rosen [1975a, 11975b, 1976], Lomet [1975], and Barth [1977]. Solutions to Excercises 1.5 and 1.6 can be found in Walter [1976] and Barth [1977].

Chapter 2

PRELIMINARY BACKGROUND

In this chapter we briefly introduce most of the jargon, notation, and basic concepts of mathematics and computer science that are employed throughout this book to discuss pre-execution analysis of computer programs. Section 2.1, Basic Mathematical Concepts and Notation, contains a cursory treatment of logic, proof techniques, sets, relations, functions, and directed graphs. Section 2.2 mentions some basic notions about algorithms such as time complexity and data structures, and then introduces Pidgin SIMPL, the language that we use to express algorithms. Section 2.3, Introduction to Lattice Theory, presents some (not so familiar but needed) concepts such as semilattices, a basic fixed point algorithm, and properties of operations on semilattices.

Much of this material is quite standard, and is included for completeness. In fact, we invite the confident reader to skip or just skim Sections 2.1 and 2.2, and begin with Section 2.3. For the skipper, needed definitions can be easily found with the aid of the Index.

2.1 Basic Mathematical Concepts and Notation

2.1.1 Logic

A *proposition* is a statement or sentence that is *true* (denoted by T) or *false* (denoted by F) but not both. T and F are called *truth values*. The words *observation, lemma, theorem,* and *corollary* denote propositions. An observation is a simple result that usually follows immediately from definitions; a lemma is a preliminary or preparatory result to a theorem; a theorem is a main result; and a corollary is a simple result, usually an immediate consequence of a theorem.

A *propositional connective* is a symbol that can be used to create compound propositions out of constituent propositions such that the truth value of the compound proposition is determined by the truth values of its constituent propositions and the definitions of the propositional connectives. If P and Q are propositions, then so are $\neg P$, $P \wedge Q$, $P \vee Q$, $P \rightarrow Q$, and $P \leftrightarrow Q$. The names and definitions of each of these propositional connectives is given in Figure 2.1.

Basic Mathematical Concepts and Notation

Symbol	Name
$\neg$	not
$\wedge$	and
$\vee$	(inclusive) or
$\rightarrow$	implies
$\leftrightarrow$	if and only if

(i)

P	Q	$\neg P$	$P \wedge Q$	$P \vee Q$	$P \rightarrow Q$	$P \leftrightarrow Q$
T	T	F	T	T	T	T
T	F	F	F	T	F	F
F	T	T	F	T	T	F
F	F	T	F	F	T	T

(ii)

Figure 2.1 Some propositional connectives.

To avoid ambiguity in evaluating compound propositions such as $P \wedge Q \rightarrow R$, we shall sometimes use parentheses and square brackets. When these are omitted, the priority (or precedence) of propositional connectives is that indicated by the order in Figure 2.1(i). That is, unless otherwise indicated by parentheses, $\neg$ "grabs first and grabs as little as possible", then $\wedge$, then $\vee$, then $\rightarrow$, then $\leftrightarrow$. Thus, $P \wedge Q \rightarrow R$ is $(P \wedge Q) \rightarrow R$, and $P \rightarrow Q \leftrightarrow \neg Q \rightarrow \neg P$ is $(P \rightarrow Q) \leftrightarrow ((\neg Q) \rightarrow (\neg P))$. When a proposition contains only one type of propositional connective, we parenthesize from the left. Thus, $P \rightarrow Q \rightarrow R$ is $(P \rightarrow Q) \rightarrow R$.

The following are equivalent:

(a) $P \rightarrow Q$.
(b) P implies Q.
(c) If P, then Q.
(d) P only if Q.
(e) Q if P.
(f) Q when P.
(g) Q is a necessary condition for P.
(h) P is a sufficient condition for Q.
(i) $\neg Q \rightarrow \neg P$.

The following are equivalent:

(a) $P \leftrightarrow Q$.
(b) $(P \rightarrow Q) \wedge (Q \rightarrow P)$.

(c) P if and only if Q.
(d) P iff Q.
(e) P is equivalent to Q.
(f) P is a characterization of Q.
(g) (P implies Q) and conversely.
(h) (P only if Q) and (P if Q).

A *definition* is an abbreviation, and is usually presented by employing 'iff'. We shall italicize any term or phrase being defined, and typically place it on the left side of an 'iff'. Since a definition is an abbreviation, its right side may be substituted for its left side and vice versa. Some definitions give convenient names to things and have a form such as 'A *this* is a that.' or 'A this is called a *that*.' We assume that the reader is familiar with the concept of a *recursive definition*.

A propositional connective, say $*$, is called *commutative* iff $P*Q$ is equivalent to $Q*P$ for all propositions P, Q; $*$ is called *associative* iff $(P*Q)*R$ is equivalent to $P*(Q*R)$ for all propositions P, Q, R. Note that $\wedge$, $\vee$, and $\leftrightarrow$ are commutative and associative, but $\rightarrow$ is neither commutative nor associative.

The *converse* of $P \rightarrow Q$ is $Q \rightarrow P$, and the *contrapositive* of $P \rightarrow Q$ is $\neg Q \rightarrow \neg P$. We call P the *hypothesis* of $P \rightarrow Q$, and Q its *conclusion*. $P_1, \ldots, P_k$ are called the *hypotheses* of $(P_1 \wedge \cdots \wedge P_k) \rightarrow Q$. Since $P \rightarrow (Q \rightarrow R)$ is equivalent to $(P \wedge Q) \rightarrow R$, P and Q are the hypotheses of $P \rightarrow (Q \rightarrow R)$.

A *predicate* is a proposition involving one or more variables where a variable essentially plays the role of a pronoun. For example, $P(x)$ is typically used to denote a predicate with one variable.

To economize writing, we shall use the symbol $\forall$ to mean 'for all', and the symbol $\exists$ to mean 'there exists'. These symbols are called *quantifiers*, and they refer to a particular (though usually tacit) universe of discourse. If $P(x)$ is a predicate involving x, then

$$\forall x P(x) \leftrightarrow \neg \exists x \neg P(x),$$

$$\neg \forall x P(x) \leftrightarrow \exists x \neg P(x),$$

$$\forall x \neg P(x) \leftrightarrow \neg \exists x P(x),$$

$$\neg \forall x \neg P(x) \leftrightarrow \exists x P(x),$$

where both $\forall$ and $\exists$ have the same precedence as $\neg$. Thus, $\neg \forall x \neg P(x)$ is $\neg (\forall x [\neg P(x)])$, and $\forall x \forall y Q(x,y)$ is $\forall x [\forall y Q(x,y)]$.

We shall sometimes distinguish between the notions of *use* and *mention* by delimiting with apostrophes a string of one or more symbols that is mentioned.

2.1.2 Proof Techniques

The words 'show' and 'prove' are synonyms.

There are several alternative methods of showing that $P \rightarrow Q$ is true. In a *direct proof*, we first assume that P is true, and then we employ P and definitions and known results to eventually establish Q. (The justification for assuming P is true is as follows. Since P is a proposition, it is either true or false but not both. If P is false, then $P \rightarrow Q$ is true by Figure 2.1(ii). So without loss of generality we can assume that P is true.) Another way to show that $P \rightarrow Q$ is true is to show that $P \rightarrow Q_1, Q_1 \rightarrow Q_2, \ldots, Q_k \rightarrow Q$. In a general *proof by contradiction* of a proposition R, we show by direct proof that $\neg R \rightarrow (S \wedge \neg S)$, for any proposition S. So, in a proof by contradiction of $P \rightarrow Q$, we show that $\neg(P \rightarrow Q) \rightarrow (S \wedge \neg S)$; that is, $(P \wedge \neg Q) \rightarrow (S \wedge \neg S)$. A *proof by contraposition* of $P \rightarrow Q$ is just a direct proof of $\neg Q \rightarrow \neg P$. In a *proof by cases* of $P \rightarrow Q$, if P is $P_1 \vee \cdots \vee P_k$ for $k \geqslant 2$, we show that for each i, $1 \leqslant i \leqslant k$, $P_i \rightarrow Q$.

To show that $P \leftrightarrow Q$ is true, we usually show that both $P \rightarrow Q$ (the "only if" part) and $Q \rightarrow P$ (the "if" part) are true. Another way to show that $P \leftrightarrow Q$ is true is to show that $P \leftrightarrow Q_1, Q_1 \leftrightarrow Q_2, \ldots, Q_k \leftrightarrow Q$.

One way to prove that $\forall x P(x)$ is true is to let x represent an arbitrary element of the universe of discourse and prove that $P(x)$ is true. (Then, since $P(x)$ is true for an arbitrary x, $P(x)$ must be true for all x; i.e., $\forall x P(x)$ is true.)

$\exists ! x P(x)$ means that there exists a unique x such that $P(x)$ is true. Since $\exists ! x P(x) \leftrightarrow [\exists x P(x) \wedge \forall y \forall z [(P(y) \wedge P(z)) \rightarrow y = z]]$, to prove that $\exists ! x P(x)$ is true we prove that both $\exists x P(x)$ (the "existence" part) and $\forall y \forall z [(P(y) \wedge P(z)) \rightarrow y = z]$ (the "uniqueness" part) are ture. The popular way of showing the uniqueness part is a proof by contradiction.

One possible formulation of mathematical induction is as follows. Let $\{0, 1, 2, \ldots\}$ be the set of *natural numbers*, and let $P(n)$ be a predicate associated with the natural numbers. To show that $P(n)$ is true for all natural numbers n, $n \geqslant k$, where k is a fixed natural number (usually 0 or 1), it suffices to show both the following:

(a) $P(k)$ is true. (This is called the *basis*.)
(b) Let $m > k$. If $P(k), P(k+1), \ldots, P(m-1)$ are true, then $P(m)$ is true. (This is called the *inductive step*.)

The above technique is called *strong mathematical induction*. Sometimes, when performing the inductive step, we can show that $P(m)$ is true by using only $P(m-1)$. This (slight simplification) is called *weak mathematical induction*.

To perform a proof by mathematical induction, we must first identify the *inductive hypothesis*; that is, $P(n)$. Also we must identify the *induction parameter*, the variable on which we shall perform induction, as well as the basis value of this parameter.

2.1.3 Sets

We postulate the existence of two distinct types of "objects": *atoms* and *sets*. We also postulate an abstract notion of *membership* between objects. If s is a member of S, we write $s \in S$. The negation of $s \in S$ is $s \notin S$. Atoms cannot have members, but sets can. If S is a set, then its *members* or *elements* are those objects s (not necessarily atoms) such that $s \in S$. We reserve the symbol $\emptyset$ for the *empty set*, the set with no members. We assume that each element of a set appears exactly once in that set, and that the order of occurrence of elements is arbitrary or undefined.

When a set S has m elements, we write $\#S = m$. If S has a finite number of elements, then we often write $S = \{s_1, s_2, \ldots, s_m\}$, where $s_1, s_2, \ldots, s_m$ are all the elements of S, with $s_i \neq s_j$ for $i \neq j$. Frequently we define a set by a predicate $P(x)$, where the set consists of exactly those elements x for which the predicate is true. Such a set is denoted by $\{x|P(x)\}$.

Let A and B be sets. A is a *subset* of B, written $A \subseteq B$, iff $\forall x[x \in A \rightarrow x \in B]$. A is *equal* to B, written $A = B$, if and only if $A \subseteq B$ and $B \subseteq A$. A is a *proper subset* of B, written $A \subsetneq B$, iff $(A \subseteq B) \wedge (A \neq B)$.

The *union, intersection,* and *difference* of A and B, respectively denoted by $A \cup B$, $A \cap B$, and $A - B$, are defined as follows:

$$A \cup B = \{x | x \in A \vee x \in B\}$$

$$A \cap B = \{x | x \in A \wedge x \in B\}$$

$$A - B = \{x | x \in A \wedge x \notin B\}$$

Let U be a set of elements (or atoms) under consideration, the *universal set*, and let $A \subseteq U$. $U - A$ is often written $\overline{A}$ and is called the *complement* of A.

Sets A and B are said to be *disjoint* iff $A \cap B = \emptyset$.

The *power set* of set A, written 2^A, is $\{B | B \subseteq A\}$.

If I is some (indexing) set such that A_i is a known set for each i in I, then $\bigcup_{i \in I} A_i = \{x | \exists i [i \in I \wedge x \in A_i]\}$.

Let s and t be (not necessarily distinct) objects. Then (s, t) denotes the *ordered pair* consisting of s and t. The ordering of the two objects is important. Two ordered pairs (s, t) and (u, v) are *equal*, written $(s, t) = (u, v)$, if and only if $s = u$ and $t = v$.

Basic Mathematical Concepts and Notation 33

For $n \geq 3$, an *ordered n-tuple* is an ordered pair whose first object is an ordered $(n-1)$-tuple. For convenience we shall always omit all parentheses of an ordered n-tuple except the outermost pair. A 3-tuple is called a *triple*, and a 4-tuple is called a *quadruple*.

The *Cartesian product* of two sets A and B, denoted by $A \times B$, is $\{(a,b) | a \in A \wedge b \in B\}$.

Sometimes we use "restricted quantifiers". If $P(x)$ is a predicate involving variable x, $Q(x,y)$ is a predicate involving variables x and y, and S is a set, then

$$(\forall x \in S) P(x) \leftrightarrow \forall x [x \in S \rightarrow P(x)],$$

$$(\exists x \in S) P(x) \leftrightarrow \exists x [x \in S \wedge P(x)],$$

$$(\forall x, y \in S) Q(x, y) \leftrightarrow \forall x \forall y [x \in S \wedge y \in S \rightarrow Q(x, y)].$$

2.1.4 Relations, Functions

A *relation R from* a set A *to* a set B is any subset of $A \times B$: $R \subseteq A \times B$. The set A is called the *domain* of R, and the set B the *range* of R. We use aRb to denote $(a,b) \in R$, and $a\slashed{R}b$ to denote $(a,b) \notin R$.

Let S be a set. A *relation R on S* is any subset of $S \times S$. The inverse of R, denoted by R^{-1}, is $\{(y,x) | xRy\}$. R is

(a) *reflexive* iff $(\forall x \in S)[xRx]$;
(b) *irreflexive* iff $(\forall x \in S)[x\slashed{R}x]$;
(c) *symmetric* iff $R = R^{-1}$;
(d) *antisymmetric* iff $(\forall x, y \in S)[xRy \wedge yRx \rightarrow x = y]$;
(e) *asymmetric* iff $(\forall x, y \in S)[xRy \rightarrow y\slashed{R}x]$;
(f) *transitive* iff $(\forall x, y, z \in S)[xRy \wedge yRz \rightarrow xRz]$.

OBSERVATION 2.1 *Let R be a relation on set S.*

(a) *Reflexivity, irreflexivity, symmetry, antisymmetry, asymmetry, transitivity are each "preserved" by inversion. For example, this means that if R is transitive, then so is R^{-1}.*
(b) *If R is both irreflexive and transitive, then R is asymmetric.*
(c) *If R is asymmetric, then R is antisymmetric.*

If Q and R are relations on S, then the *composition* of Q and R, denoted by QR, is $\{(x,z) | (\exists y \in S)[xRy \wedge yQz]\}$. The *reflexive closure* of R, denoted by $R^{\#}$, is $R \cup \{(x,x)|x \in S\}$. The *k-fold product* of R, denoted by R^k, is defined by: (a) $R^1 = R$, and (b) $R^i = RR^{i-1}$ for $i \geq 2$. The *transitive closure* of R, denoted by R^+, is $\{(x,y)|\exists i[i \geq 1 \wedge xR^iy]\}$. The *reflexive and transitive closure* of R, denoted by R^*, is $R^{\#} \cup R^+$. The *completion* of R, denoted by $\hat{R}$, is $\{(x,y)|xR^*y \wedge \neg \exists z[yRz]\}$.

An *equivalence relation* is a relation on a set that is reflexive, symmetric, and transitive. A *partition* of a set S is a set $\{B_1, B_2, \ldots, B_k\}$ such that

(a) $B_i \neq \emptyset$ for $1 \leq i \leq k$;
(b) $B_i \cap B_j = \emptyset$ for $i \leq i \neq j \leq k$; and
(c) $B_1 \cup B_2 \cup \cdots \cup B_k = S$.

An equivalence relation R on a set S partitions S into disjoint subsets called *equivalence classes*.

A *function* (AKA *mapping, transformation*) f from a set A to a set B, denoted by $f: A \to B$, is a relation from A to B such that for each $x \in A$ there is a unique $y \in B$ for which $(x,y) \in f$. $f(x) = y$ means $(x,y) \in f$.

A function $f: A \to B$ is called

(a) an *injection* iff $(\forall x, y \in A)[x \neq y \to f(x) \neq f(y)]$;
(b) a *surjection* iff $(\forall y \in B)(\exists x \in A)[f(x) = y]$; and
(c) a *bijection* (or *one-to-one correspondence*) iff f is both an injection and a surjection.

The *composition* of two functions $f: B \to C$ and $g: A \to B$, denoted by fg, is $\{(x,z)|(\exists y \in B)[g(x) = y \wedge f(y) = x]\}$. Thus, $fg(x) = f(g(x))$.

Let A be a set. A^n is defined by: (a) $A^1 = A$, and (b) $A^i = A \times A^{i-1}$ for $i \geq 2$. A function from A^n to A, for $n \geq 1$, is called an *n-ary operation* on A. *Unary* and *binary* are synonyms for 1-ary and 2-ary. The *identity operation* on A is $\{(x,x)|x \in A\}$.

A *fixed point* of a unary operation f on a set A is an element x of A such that $f(x) = x$.

Intuitively, the *cardinality* of a set A is the number of elements in A. Formally, two sets A and B are of *equal cardinality* iff there exists a bijection f from A to B. A set A is called *infinite* iff it is of equal cardinality with a proper subset of itself, and finite otherwise. When a set is finite, we can assign a natural number to its cardinality (or size). A set is called *denumerable* (or *countably infinite*) iff it has the same cardinality as the set of natural numbers, and *countable* iff it is either finite or denumerable.

2.1.5 Directed Graphs

A *directed graph* (or *graph*) G is a pair (N, A), where N is a finite nonempty set, and A is a relation on N. Each element in N is called a *node*,

Basic Mathematical Concepts and Notation 35

and each pair in A is called an *arc*. Throughout this book, $n = \#N$ and $r = \#A$.

The arc (x,y) *leaves* the *tail* node x and *enters* the *head* node y. We say that x is a *predecessor* of y, and y is a *successor* of x. PRED(x) denotes the set of predecessors of node x, and SUC(x) denotes the set of successors of node x. The *in-degree* of a node x is $\#\text{PRED}(x)$, and the *out-degree* of x is $\#\text{SUC}(x)$.

We shall draw graphs by using circles (and sometimes boxes) to denote nodes, and arrows between nodes to denote arcs. Each node of a graph is usually assigned a unique number from 1 to n that is drawn inside the node to identify it.

Let $G = (N, A)$ be a directed graph. A *node labeling* of G is a function $f: N \to L$ that maps each node to an element of set L. Node labels are drawn outside but adjacent to a node. Similarly, an *arc labeling* of G is a function $g: A \to M$ that maps each arc to an element of set M.

A *path* is a finite sequence of one or more nodes such that, if the sequence contains two or more nodes and is denoted by $(x_1, \ldots, x_k)$, $k \geq 2$, then $(x_i, x_{i+1}) \in A$ for $1 \leq i \leq k-1$. A *simple path* is a path with distinct nodes. If $k \geq 2$, then $(x_1, \ldots, x_k)$ is a path *from* x_1 *to* x_k with *length* $k-1$, and x_k is said to be *accessible* from x_1. As a special case, a single node denotes a path of length 0 from itself to itself. Occasionally, a path is represented by its sequence of arcs $((x_1, x_2), \ldots, (x_{k-1}, x_k))$. A *cycle* is a path $(x_1, \ldots, x_k)$ in which $x_1 = x_k$. A *simple cycle* is a cycle from a node to itself. A path is *cycle-free* (or *acyclic*) iff it does not contain a cycle.

A *directed acyclic graph* (acronym: *DAG*) is a directed graph that has no cycles.

Let $G = (N, A)$ be a graph, and let $A' \subseteq A$. The *subgraph generated by A'* (i.e., a *generated subgraph*) is a graph $G' = (N', A')$, where $N' \subseteq N$ and $A' = A \cap (N' \times N')$. A *partial graph* of G is a graph $G'' = (N, A'')$, where $A'' \subseteq A$. Thus a *partial subgraph generated by A'* is a graph $G''' = (N', A''')$, where $A''' \subseteq A'$.

Let $G = (N, A)$ be a graph. We can partition N into equivalence classes N_i, $1 \leq i \leq k$, such that nodes x and y are in the same equivalence class if and only if there is a path from x to y and from y to x. Let $A_i = A \cap (N_i \times N_i)$, for $1 \leq i \leq k$. Each generated subgraph $G_i = (N_i, A_i)$, $1 \leq i \leq k$, is called a *strongly connected component* (acronym: *SCC*) of G. A graph is called *strongly connected* iff it has exactly one strongly connected component. The *acyclic condensation* of G is a DAG G'', the nodes of which are the SCCs of G, denoted by SCC_i, and the arcs of G'' satisfy: arc $(\text{SCC}_i, \text{SCC}_j)$ is in G'' if and only if there exists an arc (u, v) in G such that u is in N_i and v is in N_j.

A (rooted) *tree* is a DAG satisfying the following properties:

(a) There is exactly one node, called the *root*, with in-degreee 0.
(b) Every node except the root has in-degree 1.
(c) There is a (unique) path from the root to each node.

If (u,v) is an arc in a tree, then u is called the *parent* of v, and v is called a *child* of u. The *ancestor* and *descendant* relations are the respective reflexive and transitive closures of the parent and child relations. Node x is called a *proper ancestor* of node y iff x is an ancestor of y and $x \neq y$. The *proper descendant* relation is defined analogously.

An *ordered tree* is a tree in which the children of each node are ordered from left to right ("oldest to youngest").

2.2 Comments about Algorithms

2.2.1 Algorithms

Algorithms are used to solve certain well defined "problems". An *algorithm* is a finite sequence of "instructions", where (a) each individual instruction can be executed mechanically in a fixed amount of time with a fixed amount of effort; (b) there are zero or more "inputs" and one or more "outputs"; and (c) execution of the algorithm always halts (terminates, stops) after a finite amount of time for all (suitable) inputs.

The *time complexity* of an algorithm is the aggregate amount of time, expressed as a function of the size of the input, needed by the algorithm to solve that problem. The limiting behavior of the time complexity as the size of a problem increases is called the *asymptotic time complexity*. The definitions for *space complexity* and *asymptotic space complexity* are analogous. We shall normally use *worst-case complexity*, the maximum complexity over all inputs of a given size.

We shall use *big-Oh* (or *order-of-magnitude*) *notation* to express the asymptotic time and space complexity of an algorithm. If an algorithm processes inputs of size n in time cn^2 for some constant c, then we express the time complexity of that algorithm as $O(n^2)$, read "order n^2". More precisely, a function $g(n)$ is said to be $O(f(n))$ iff there exists a constant c such that $g(n) \leq cf(n)$ for all but some finite (possibly empty) set of nonnegative values n.

We shall express the time complexity of an algorithm by counting "steps". Thus, we need to specify a formal model of computation in order to define what is meant by a basic step in an algorithm. Here we (tacitly) assume that the computing device for executing algorithms is a *random access machine* (acronym: *RAM*). A RAM is a particular formal model of computation with assembly-language-like instructions, which has been well studied elsewhere.[1]

To specify time and space complexity exactly, we must specify the time required to execute each RAM instruction and the space used by each RAM "word". Unless otherwise mentioned, we use the *uniform cost criterion* in which each RAM instruction requires one unit of time (one step) and each word requires one unit of space. The time complexity of a RAM program is not always realistic, because words are of unbounded size. The *logarithmic cost criterion*, another possible cost measure, is based on the assumption that the cost of performing an instruction is proportional to the sum of the lengths of the operands of the instruction, and that[2] $\lceil \log(n+1) \rceil$ bits are required to represent the nonnegative interger n.

For computations involving ordinary integer arithmetic and pointer operations in data structures we shall use the term *elementary step* when treating time complexity. For logical operations such as AND, OR, NOT on bit vectors we use the term *bit vector step* or *extended step*. Thus, a logical operation on bit vectors of length m requires m elementary steps but only 1 bit vector step.

A "problem" is called *decidable* iff there exists an algorithm to solve it, and *undecidable* otherwise. In Chapter 9 we shall prove that the "meet over all paths" problem for monotone frameworks is undecidable.

We assume that the reader is familiar with the concept of a *recursive algorithm*.

2.2.2 Data Structures

We assume that the reader is already familiar with the basic data structures for sequences, sets, functions, and directed graphs. Here, we just review the names of a few that are mentioned in the sequel.

Data structures for sets include *singly linked lists, doubly linked lists, hash tables,* and *bit vectors*, among many others. When using a bit vector representation for sets, we assume that the universe of discourse U (of which all sets are subsets) has m members, and that each atom of U is assigned a unique bit position i, for $1 \leqslant i \leqslant m$. A subset S of U is represented by a vector of m bits, where the ith bit is 1 iff the ith atom of U is in S, and 0 otherwise. Typical efficient operations on sets represented by bit vectors include membership interrogation for an atom, union, intersection, complement, and equality testing.

---
[1] See Aho, Hopcroft, and Ullman [1974], for example.
[2] Throughout this book, all logarithms are to the base 2. $\lceil x \rceil$, the *ceiling* of x, denotes the smallest integer greater than or equal to x.

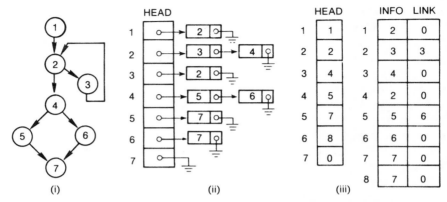

Figure 2.2 Some representations of a directed graph. (i) Pictorial representation; (ii) successor lists representation; (iii) tabular representation of successor lists.

Data structures for functions include hash tables and (the ubiquitous) arrays. For example, hashing can be used to "look up" the index associated with an atom when performing membership interrogation with bit vectors. If the nodes of a graph are numbered from 1 to n, then a node labeling is easily represented by an array.

Among the data structures we shall use for directed graphs are *successor lists* and *predecessor lists*. Successor lists, also called *adjacency lists*, have a singly linked list for each set SUC(x). An example of this data structure is given in Figure 2.2. Predecessor lists have a singly linked list for each set PRED(x).

2.2.3 Pidgin SIMPL

We shall present algorithms in a high-level language called *Pidgin* SIMPL[3] that uses traditional programming language constructs such as procedures, functions, statements, expressions, and conditions. Usually we package an algorithm into a procedure and describe the input and output in the accompanying text rather than using formal parameters. For convenience, whenever the word 'procedure' occurs below, we take it to mean either a procedure or a function.

A Pidgin SIMPL *program* consists of a (possibly empty) sequence of global declarations, followed by a sequence of one or more (unnested though possibly recursive) procedures, followed by a statement indicating the procedure with which execution is to begin. All procedures can access

---

[3]No, the E is not missing from SIMPL. For a complete description of SIMPL (sans pigeons) see Basili and Turner [1975]. We give a synopsis of SIMPL-T in Section 10.2.3.

Comments about Algorithms

global variables. Any procedure may have local variables. For our needs in this book, each formal parameter of a procedure that is a scalar data type (e.g., integer) is assumed to be call-by-value, and each formal parameter of a procedure that is a structure data type (e.g., array) is assumed to be call-by-reference.

Some statements in Pidgin SIMPL include the following:

(a) variable := expression
(b) **if**[4] condition **then** statement(s) **endif**
(c) **if** condition **then** statement(s)1 **else** statement(s)2 **endif**
(d) **while** condition **do** statement(s) **endwhile**
(e) **repeat** statement(s) **until** condition
(f) **for** variable := initial-value **to** final-value **by** step-size **do** statement(s) **endfor**
(g) **call** procedure-name(actual-parameter-list)
(h) **procedure** procedure-name(formal-parameter-list) statement(s) **return**
(i) **function** function-name(formal-parameter-list) statement(s) **return** (expression)

There are three types of loops in Pidgin SIMPL: **while, repeat,** and **for**. (SIMPL-T only has a **while**-loop.) In (h) and (i) each formal parameter is preceded by its type.

Pidgin SIMPL has no block structure. There is no statement terminator nor separator such as the semicolon. Comments are delimited by '/*' and '*/'. We shall use a simple (self-evident) indentation style to enhance the readability of statements and procedures. The dollar sign '$' can be used in identifiers.

We frequently use English sentences in place of equivalent sequences of (atomic) programming language statements. This allows us to submerge easy implementation details and to present algorithms in a more readable fashion. We also employ some high-level types in declarations such as sets and directed graphs when convenient.

Input and output will generally be described in English. Whenever local or global variables are not declared in the program, either they will be described in the accompanying text or their role will be obvious.

Whenever we use English statements or high-level data types in Pidgin SIMPL, there is a straightforward way to translate them into a RAM program such that the asymptotic time and space complexities of the Pidgin SIMPL program and the RAM program are equivalent.

---

[4]Keywords in Pidgin SIMPL are denoted by lower case boldface. In Chapter 10 we use upper case boldface for keywords of SIMPL-T.

2.3 Introduction to Lattice Theory

Many concepts in this book come from an area of mathematics called *lattice theory*. For example, we shall talk about partially ordered sets, topological sorting, semilattices, and monotonic operations on semilattices.

2.3.1 Posets, Chains

A *partial ordering* on a set S, which we denote by the symbol $\sqsubseteq$, is a reflexive, antisymmetric, and transitive relation on S. The pair $(S, \sqsubseteq)$ is called a *poset* (short for *partially ordered set*). We write $x \sqsubset y$ if and only if $x \sqsubseteq y$ and $x \neq y$. In addition, $\sqsubset$ is called the *reflexive reduction* of $\sqsubseteq$ because $\sqsubset = \sqsubseteq - \{(x,x) | x \in S\}$. We sometimes write $y \sqsupseteq x$ for $x \sqsubseteq y$, and $y \sqsupset x$ for $x \sqsubset y$.

OBSERVATION 2.2 *If $(S, \sqsubseteq)$ is a poset, then so is $(S, \sqsupseteq)$. Note that $\sqsupseteq$ is the inverse of $\sqsubseteq$; that is, $\sqsubseteq^{-1}$.*

OBSERVATION 2.3 *If $(S, \sqsubseteq)$ is a poset, then $\sqsubset$ is irreflexive and transitive. Conversely, if $\sqsubset$ is an irreflexive and transitive relation on S and $x \sqsubseteq y$ is defined by $x \sqsubset y$ or $x = y$, then $(S, \sqsubseteq)$ is a poset.*

OBSERVATION 2.4 *If $(S, \sqsubseteq)$ is a poset and $T \subseteq S$, then[5] $(T, \sqsubseteq \cap T^2)$ is a poset.*

Let $(S, \sqsubseteq)$ be a poset, and let $T \subseteq S$. A *minimal* element of T is an element $a \in T$ such that there is no $x \in T$ for which $x \sqsubset a$. A *maximal* element of T is an element $b \in T$ such that there is no $x \in T$ for which $b \sqsubset x$. The *zero* (AKA *bottom, minimum, least*) element of S is an element $\mathbf{0} \in S$ such that $\mathbf{0} \sqsubseteq x$ for all $x \in S$. The *one* (AKA *top, maximum, greatest*) element of S is an element $\mathbf{1} \in S$ such that $x \sqsubseteq \mathbf{1}$ for all $x \in S$. A poset may not have a zero or a one.

Example 2.1 Two posets are shown in Figure 2.3. Whenever there is a line from x to y and x is pictured as lower than y, then $x \sqsubseteq y$. Transitive lines are omitted for clarity. The poset in Figure 2.3(i) has three maximal elements, two minimal elements, no zero, and no one. The poset in Figure 2.3(ii) has a one element g and a zero element p.

Elements x and y of a poset $(S, \sqsubseteq)$ are called *comparable* if either $x \sqsubseteq y$ or $y \sqsubseteq x$, and *incomparable* otherwise.

---

[5]Don't be confused by $\sqsubseteq \cap T^2$. Since T is a set, T^2 is $T \times T$. Since $\sqsubseteq$ and $T \times T$ are both relations and hence sets, they can be intersected. $\sqsubseteq \cap T^2$ is called "the restriction of $\sqsubseteq$ to T".

Introduction to Lattice Theory

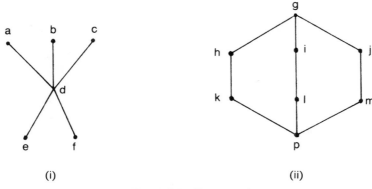

Figure 2.3 Two posets.

A *chain* (or *linear ordering*) on a set S is a partial ordering on S such that every pair of elements is comparable. Sometimes we use $s_1 \sqsubset s_2 \sqsubset \cdots \sqsubset s_k$ to denote a chain on $S = \{s_1, s_2, \ldots, s_k\}$. The *length* of $s_1 \sqsubset s_2 \sqsubset \cdots \sqsubset s_k$ is k. We are particularly interested in posets all of whose chains are of finite length.

Example 2.2 In Figure 2.3(ii), $p \sqsubset k \sqsubset h \sqsubset g$ is a chain of length four. Elements k and m in this figure are incomparable.

OBSERVATION 2.5 *Any finite nonempty subset of a poset has minimal and maximal elements. Any finite nonempty chain has unique minimum and maximum elements.*

OBSERVATION 2.6 *Let $(S, \sqsubset)$ be a poset. Define a relation $\sqsubset_n$ on S^n by $(s_1, \ldots, s_n) \sqsubset_n (t_1, \ldots, t_n)$ iff $s_i \sqsubset t_i$ for all i, $1 \leq i \leq n$. Then $(S^n, \sqsubset_n)$ is a poset. Moreover, if $\mathbf{0}$ is the zero element of $(S, \sqsubset)$, then $(\mathbf{0}, \ldots, \mathbf{0})$ is the zero element of $(S^n, \sqsubset_n)$; if $\mathbf{1}$ is the one element of $(S, \sqsubset)$, then $(\mathbf{1}, \ldots, \mathbf{1})$ is the one element of $(S^n, \sqsubset_n)$.*

OBSERVATION 2.7 *Let $(S, \sqsubset)$ be a poset all of whose chains are of finite length. Then all chains of $(S^n, \sqsubset_n)$ are also of finite length.*

2.3.2 Topological Sorting

Let $(S, \sqsubset)$ be a poset. The *reflexive and transitive reduction* of $\sqsubset$ is $\sqsubset - \{(x, y) \mid \exists i [i \geq 2 \text{ and } x \sqsubset^i y]\}$. Intuitively, the reflexive and transitive reduction of $\sqsubset$ is $\sqsubset$ minus all transitive lines. When S is finite, both $\sqsubset$ and the reflexive and transitive reduction of $\sqsubset$ are DAGs.

Let R be a partial ordering on set S, and let T be a linear ordering on set S. Then T *topologically sorts* (or *topsorts*) R iff $R \subseteq T$.

Intuitively, topological sorting corresponds to squeezing a DAG into a single column of nodes such that all arcs point downward. The position of the nodes in the column yields the linear ordering.

Here is the (hopefully) familiar algorithm for topological sorting.[6]

ALGORITHM 2.1 *Topological sorting.*

INPUT: Either the graph of $\sqsubset$ or the graph of the reflexive and transitive reduction of $\sqsubset$, where $(S,\sqsubset)$ is a poset and $\#S = n \geq 1$. The input graph is represented by successor lists and a node labeling. Initially, the label of a node is the in-degree of that node. As execution of the algorithm progresses, the label of a node is its "current" in-degree.

OUTPUT: A linear ordering $s_1 \sqsubset s_2 \sqsubset \cdots \sqsubset s_n$ on S, where $s_i \in S$ for $1 \leq i \leq n$.

METHOD: See procedure TOPSORT in Figure 2.4. $\square$

> **procedure** TOPSORT(DAG G)
>   **set** Q /* This local declaration declares Q a set */
>   Initialize Q to the set of all nodes in G with in-degree 0.
>   **while** $Q \neq \emptyset$ **do**
>   Select any node x from Q and output x.
>   Erase all arcs that leave x, and delete x from Q.
>   Add to Q all nodes whose in-degree just became 0.
>   **endwhile**
>   **return**

Figure 2.4 Algorithm for topological sorting.

Algorithm 2.1 takes $O(\max(n,r))$ elementary steps, and $O(\max(n,r))$ space. (Recall that $n = \#N$ and $r = \#A$ throughout this book.)

Example 2.3 Consider the partial ordering depicted by its reflexive and transitive reduction in Figure 2.5(i). Since node 1 has no entering arcs, initially $Q = \{1\}$. Since Q is nonempty, we output 1, erase arc (1,2), delete 1 from Q, and add 2 to Q. Next, output 2, erase arcs (2,3), (2,4), and (2,5), delete 2 from Q, and add 3,4,5 to Q. Next, either 3 or 4 or 5 can be selected. One of several possible topsorts of Figure 2.5(i) is 1,2,3,4,5,6,7,8,9,10.

The observation below is used in Chapter 5.

OBSERVATION 2.8 *If Q and R are partial orderings on set S, $Q \subseteq R$, and linear ordering T (on S) topsorts R, then T topsorts Q.*

---

[6]See Knuth [1974], for example.

Introduction to Lattice Theory

Proof Since T topsorts R, $R \subseteq T$. $Q \subseteq R$ and $R \subseteq T$ together imply $Q \subseteq T$. Thus, T topsorts Q. □

Example 2.4 Let Q and R be the partial orderings depicted by their reflexive and transitive reductions in Figures 2.5(i) and (ii) respectively. Q is a subset of R. By observation 2.8, consequently, any linear ordering that topsorts Figure 2.5(ii) also topsorts Figure 2.5(i). In particular, consider the one presented in Example 2.3.

2.3.3 Lattices, Semilattices, Boolean Algebras

Let $(S, \sqsubseteq)$ be a poset, and let a and b be elements of S. A *join* (AKA *least upper bound, lub*) of a and b is an element $c \in S$ such that $[a \sqsubseteq c$ and $b \sqsubseteq c$ and there is no $x \in S$ such that $(a \sqsubseteq x \sqsubseteq c$ and $b \sqsubseteq x \sqsubseteq c)]$. If elements a and b in S have a unique join, it is denoted by $a \sqcup b$. A *meet* (AKA *greatest lower bound, glb*) of a and b is an element $d \in A$ such that $[d \sqsubseteq a$ and $d \sqsubseteq b$ and there is no $x \in S$ such that $(d \sqsubseteq x \sqsubseteq a$ and $d \sqsubseteq x \sqsubseteq b)]$. If elements a and b in S have a unique meet, it is denoted by $a \sqcap b$.

A *lattice* is a poset $(S, \sqsubseteq)$, any two elements of which have a unique join and meet. A lattice is sometimes denoted by the triple $(S, \sqcup, \sqcap)$. The cardinality of a lattice is the cardinality of S.

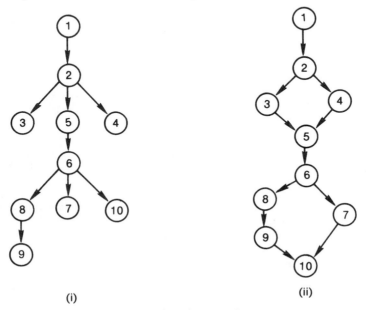

Figure 2.5 Two DAGs.

Example 2.5 If S is a finite set, then $(2^S, \cup, \cap)$ is a lattice, the "underlying" poset of which is $(2^S, \subseteq)$.

An *algebraic system* is an $(m+1)$-tuple $(S, f_1, f_2, \ldots, f_m)$, where $m \geq 1$, S is a nonempty set, and each f_i is an h_i-ary operation on S, $h_i \geq 0$. (A 0-*ary* or *nullary* operation on S is a constant element of S.)

Example 2.6 Any lattice $(S, \sqcup, \sqcap)$ is an algebraic system.

Example 2.7 A *power set algebra* $(2^T, \cup, \cap, ^-, \emptyset, T)$ is an algebraic system, where T is a finite nonempty set; union, intersection, and complement have their usual set-theoretic meanings; and $\emptyset$ is the empty set.

Let $(S, \sqsubseteq)$ be a lattice, and let $T \subseteq S$. An *upper bound* of T is an element $a \in S$ such that $(\forall x \in T)[x \sqsubseteq a]$. The *least upper bound* (AKA *lub*, *supremum*) of a subset $T \subseteq S$, denoted by $\bigsqcup T$, is an upper bound that is $\sqsubseteq$ every other upper bound. A *lower bound* of T is an element $b \in S$ such that $(\forall x \in T)[b \sqsubseteq x]$. The *greatest lower bound* (AKA *glb*, *infimum*) of a subset $T \subseteq S$, denoted by $\bigsqcap T$, is a lower bound which is $\sqsupseteq$ every other lower bound.

OBSERVATION 2.9 *Let $(S, \sqsubseteq)$ be a lattice, and let $T \subseteq S$.*

(a) $(\forall x \in T)[x \sqsubseteq \bigsqcup T]$.
(b) $(\forall x \in T)[x \sqsubseteq z] \to \bigsqcup T \sqsubseteq z$.

Similarly,

(c) $(\forall x \in T)[\bigsqcap T \sqsubseteq x]$.
(d) $(\forall x \in T)[z \sqsubseteq x] \to z \sqsubseteq \bigsqcap T$.

OBSERVATION 2.10 *The join and meet operations of a lattice $(S, \sqcup, \sqcap)$ enjoy the following properties for all $x, y, z \in S$:*

(a) $x \sqcap x = x$, $x \sqcup x = x$;   (*idempotency*).
(b) $x \sqcup y = y \sqcup x$, $x \sqcap y = y \sqcap x$;   (*commutativity*).
(c) $x \sqcup (y \sqcup z) = (x \sqcup y) \sqcup z$, $x \sqcap (y \sqcap z) = (x \sqcap y) \sqcap z$;   (*associativity*).
(d) $x \sqcup (x \sqcap y) = x$, $x \sqcap (x \sqcup y) = x$;   (*absorption*).
(e) $x \sqcup y = y$, $x \sqcap y = x$, $x \sqsubseteq y$ are equivalent.

OBSERVATION 2.11 *A set S and two binary operations $\sqcup$ and $\sqcap$ on S is a lattice if both $\sqcup$ and $\sqcap$ are idempotent, commutative, and associative.*

Introduction to Lattice Theory

By a *lattice theoretical proposition* we mean a well-formed proposition about a lattice containing some or all of the symbols $\sqsubseteq$, $\sqsupseteq$, $\sqcup$, $\sqcap$, $\bigsqcup$, $\bigsqcap$, 0, 1. The *dual* of a lattice theoretical proposition is obtained by interchanging $\sqsubseteq$ and $\sqsupseteq$, $\sqcup$ and $\sqcap$, $\bigsqcup$ and $\bigsqcap$, 0 and 1. Due to the "symmetry" of the definitions and properties of lattices, we have the following.

OBSERVATION 2.12 (The Lattice Theoretical Duality Principle) *The dual of any true lattice theoretical proposition is itself a true lattice theoretical proposition.*

Some lattices have special properties. A lattice $(S, \sqcup, \sqcap)$ is

(a) *of finite length*[7] iff each chain in the lattice is finite;
(b) *bounded* iff it has both **0** and **1**; and
(c) *distributive* iff for all $x, y, z \in S$ we have $x \sqcup (y \sqcap z) = (x \sqcup y) \sqcap (x \sqcup z)$ and $x \sqcap (y \sqcup z) = (x \sqcap y) \sqcup (x \sqcap z)$.

A *semilattice* is a pair $(S, *)$, where S is a nonempty set, and $*$ is a binary operation on S which is idempotent, commutative, and associative. If $(S, \sqcup, \sqcap)$ is a lattice, then both $(S, \sqcup)$ and $(S, \sqcap)$ are semilattices.

In a bounded lattice $(S, \sqcup, \sqcap)$, a *complement* of an element $a \in S$ is an element $b \in S$ such that $a \sqcup b = 1$ and $a \sqcap b = 0$. A bounded lattice is called *complemented* iff for each element $a \in S$ there exists a complement $b \in S$.

OBSERVATION 2.13 *In a bounded distributed lattice, complements are unique when they exist.*

We use the notation $\bar{a}$ for the unique complement of an element a.

A *Boolean algebra* is a bounded, distributive, and complemented lattice. A Boolean algebra is sometimes denoted as an algebraic system by the 6-tuple $(S, \sqcup, \sqcap, ^-, 0, 1)$, where $(S, \sqcup, \sqcap)$ is a lattice, **0** and **1** are the zero and one elements of S, and $^-$ is the unary complementation operation.

OBSERVATION 2.14 *Let $(S, \sqcup, \sqcap, ^-, 0, 1)$ be a Boolean algebra.*

(a) $(\forall x, y \in S)[\overline{x \sqcup y} = \bar{x} \sqcap \bar{y}]$.
(b) $(\forall x, y \in S)[\overline{x \sqcap y} = \bar{x} \sqcup \bar{y}]$.   DeMorgan's law
(c) $(\forall x \in S)[\bar{\bar{x}} = x]$.

---
[7] Kildall [1973] and Kam and Ullman [1975, 1976] use the term 'bounded' for 'of finite length'. Wegbreit [1975] uses the term 'well-founded'.

Two algebraic systems $(S,f_1,f_2,\ldots,f_m)$ and $(T,g_1,g_2,\ldots,g_p)$ are *isomorphic* iff $\#S = \#T$, $m = p$, f_i and g_i have the same arity for $1 \leq i \leq m$, and there exists a bijection $\psi: S \to T$ such that $(\forall i = 1, 2, \ldots, m)(\forall x_1, x_2, \ldots, x_{h_i} \in S)$

$$\psi(f_i(x_1,\ldots,x_{h_i})) = g_i(\psi(x_1),\ldots,\psi(x_{h_i})).$$

OBSERVATION 2.15 (The Stone Representation Theorem) *Every finite Boolean algebra* $(S, \sqcup, \sqcap, ^-, \mathbf{0}, \mathbf{1})$ *is isomorphic to a power set algebra* $(2^T, \cup, \cap, ^-, \emptyset, T)$.

Example 2.8 The Boolean algebra $(\{\mathbf{0},\mathbf{1}\}, \sqcup, \sqcap, ^-, \mathbf{0}, \mathbf{1})$ is isomorphic to the power set algebra $(2^{\{a\}}, \cup, \cap, ^-, \emptyset, \{a\})$ via $\psi(\mathbf{0}) = \emptyset$ and $\psi(\mathbf{1}) = \{a\}$.

2.3.4 A Fixed Point Theorem

Let $(S, \sqsubseteq)$ be a poset. An operation $f: S \to S$ on S is called *monotonic* (AKA *order-preserving, monotone, isotonic, isotone*) iff $(\forall x, y \in S)[x \sqsubseteq y \to f(x) \sqsubseteq f(y)]$.

OBSERVATION 2.16 *If f and g are monotonic operations on a poset, then so is fg. That is, composition preserves monotonicity.*

A sequence $s_0, s_1, \ldots$ in a poset $(S, \sqsubseteq)$ of finite length *converges* to an element $t \in S$ iff there exists a $k \geq 0$ such that for all $i \geq k$ we have $s_i = t$.

THEOREM 2.1 *Let $f: S \to S$ be a monotonic operation on a poset $(S, \sqsubseteq)$ with a $\mathbf{0}$ and of finite length. The least fixed point of f is $f^k(\mathbf{0})$, where $f^0(x) = x$, $f^{i+1}(x) = f(f^i(x))$ for $i \geq 0$, $f^k(\mathbf{0}) = f(f^k(\mathbf{0}))$, and there is no j, $0 \leq j < k$, such that $f^j(\mathbf{0}) = f(f^j(\mathbf{0}))$.*

Proof To show that there exists a k, $k \geq 0$, such that $f^k(\mathbf{0})$ is a fixed point, consider the sequence $\mathbf{0}, f(\mathbf{0}), f^2(\mathbf{0}), \ldots$. By monotonicity, $\mathbf{0} \sqsubseteq f(\mathbf{0}) \sqsubseteq f^2(\mathbf{0}) \sqsubseteq \cdots$. If an infinite number of these were distinct, then $(S, \sqsubseteq)$ would not be of finite length. Thus, for some k, $k \geq 0$, $f^k(\mathbf{0})$ is a fixed point, and $\mathbf{0}, f(\mathbf{0}), f^2(\mathbf{0}), \ldots$ converges to $f^k(\mathbf{0})$. That is, $\mathbf{0} \sqsubseteq f(\mathbf{0}) \sqsubseteq f^2(\mathbf{0}) \sqsubseteq \cdots \sqsubseteq f^k(\mathbf{0}) = f^{k+1}(\mathbf{0}) = \cdots$.

To show that $f^k(\mathbf{0})$ is a least fixed point, we show by induction on i that if p is any other fixed point, then $f^i(\mathbf{0}) \sqsubseteq p$.

BASIS: ($i = 0$). $\mathbf{0} \sqsubseteq p$, by the definition of $\mathbf{0}$.

INDUCTIVE STEP: ($i > 0$). Assume that $f^{i-1}(\mathbf{0}) \sqsubseteq p$, and consider the case for i. By monotonicity, $f(f^{i-1}(\mathbf{0})) = f^i(\mathbf{0}) \sqsubseteq f(p) = p$. $\square$

Theorem 2.1 gives us Algorithm 2.2.

Introduction to Lattice Theory

ALGORITHM 2.2 *The basic fixed point algorithm.*

INPUT: A monotonic operation $f: S \to S$ on a join-semilattice with a **0** (dually: meet-semilattice with a **1**) and of finite length.

OUTPUT: The least fixed point $f^k(\mathbf{0})$ (dually: greatest fixed point $f^k(\mathbf{1})$), where k is also least.

METHOD: See function ITERATE in Figure 2.6.   □

---

function ITERATE(monotonic operation f on a poset $(S, \sqsubseteq)$ with a **0** / * dually: a **1** */ and of finite length)
  poset variable x
  $x := \mathbf{0}$ /* dually: **1** */
  while $x \neq f(x)$ **do** $x := f(x)$ **endwhile**
return(x) /* $= f(x)$ */

Figure 2.6 **The basic fixed point algorithm.**

2.3.5 Properties of Operations on Semilattices

In this section we shall present some relationships among various properties of an operation on a semilattice. These properties include monotonicity, the endomorphism property, continuity, and the GKUW[8] property.

Let $(S, \sqcap)$ be a semilattice. An operation $f: S \to S$ on S

(a) is called a *meet-endomorphism* (AKA *distributivity*) iff $(\forall x, y \in S)[f(x \sqcap y) = f(x) \sqcap f(y)]$;
(b) has the *GKUW property* iff $(\forall x, y \in S)[f(x \sqcap y) \sqsubseteq f(x) \sqcap f(y)]$; and
(c) is called *continuous* iff for all directed subsets $T \subseteq S$ we have $f(\sqcup T) = \sqcup \{f(x) \mid x \in T\}$. A nonempty subset $T \subseteq S$ is called *directed* iff every nonempty finite subset of T has a lower bound belonging to T.

OBSERVATION 2.17 *An operation on a semilattice is monotonic if and only if it has the GKUW property.*

Proof

If: Assume f has the GKUW property and let $x \sqsubseteq y$. Then $x \sqcap y = x$. By the GKUW property, $f(x \sqcap y) = f(x) \sqsubseteq f(x) \sqcap f(y)$. But $f(x) \sqsubseteq f(x) \sqcap f(y)$ means $f(x) \sqcap (f(x) \sqcap f(y)) = f(x)$, or $f(x) \sqcap f(y) = f(x)$

---

[8]This property is named after Graham, Kam, Ullman, and Wegman (in alphabetical order) who have observed it.

after using associativity and idempotency. Finally, $f(x) \sqcap f(y) = f(x)$ means $f(x) \sqsubseteq f(y)$. Hence, f is monotonic.

Only if: Assume f is monotonic. By the definition of $\sqcap$, $x \sqcap y \sqsubseteq x$ and $x \sqcap y \sqsubseteq y$. By monotonicity, $f(x \sqcap y) \sqsubseteq (x)$ and $f(x \sqcap y) \sqsubseteq f(y)$. But $a \sqsubseteq b$ and $c \sqsubseteq d$ imply $a \sqcap c \sqsubseteq b \sqcap d$. Thus, $f(x \sqcap y) \sqcap f(x \sqcap y) = f(x \sqcap y) \sqsubseteq f(x) \sqcap f(y)$, by absorption. □

OBSERVATION 2.18 *If an operation on a semilattice is continuous, then it is a meet-endomorphism.*

Proof If f is a continuous operation on a semilattice $(S, \sqcap)$, then for each subset $T \subseteq S$ for which $\#T = 2$, $f(\sqcap T) = \sqcap \{f(x) | x \in T\}$. That is, $(\forall x, y \in S)[f(x \sqcap y) = f(x) \sqcap f(y)]$. □

OBSERVATION 2.19 *Any meet-endomorphism is monotonic.*

Proof Let f be a meet-endomorphism on a semilattice $(S, \sqcap)$, and let $x \sqsubseteq y$. Then $x \sqcap y = x$. So, $f(x) = f(x \sqcap y) = f(x) \sqcap f(y)$, and $f(x) = f(x) \sqcap f(y)$ means $f(x) \sqsubseteq f(y)$. Thus, f is monotonic. □

OBSERVATION 2.20 *If f is a meet-endomorphism on a semilattice $(S, \sqcap)$, then for any nonempty finite subset $T \subseteq S$, $f(\sqcap T) = \sqcap \{f(x) | x \in T\}$.*

Proof Assume the hypotheses, and let T be a nonempty finite subset of S.

INDUCTIVE HYPOTHESIS: $f(\sqcap T) = \sqcap \{f(x) | x \in T\}$, where $\#T = i$.

BASIS: $(i = 1)$. If $T = \{x\}$, then $\sqcap T = x$ and $\sqcap \{f(x)\} = f(x)$, so $f(\sqcap T) = f(x) = \sqcap \{f(x)\} = \sqcap \{f(x) | x \in T\}$.

INDUCTIVE STEP: $(i > 1)$. Assume the inductive hypothesis for $i - 1$, and consider the case for i. Let $T = T' \cup \{y\}$. Then $f(\sqcap T) = f(y \sqcap (\sqcap T')) = f(y) \sqcap f(\sqcap T') = f(y) \sqcap (\sqcap \{f(x) | x \in T'\}) = \sqcap \{f(x) | x \in T\}$, using the endomorphism property of f and the inductive hypothesis. □

Note that the property in Observation 2.20 is not necessarily continuity. Observations 2.17–2.19 are summarized in Figure 2.7.

Continuity
↓
Meet-endomorphism
↓
GKUW property ↔ Monotonicity

Figure 2.7   Relationships among properties.

Bibliographic Notes

Bittinger [1972] is a good source for logic and proof techniques. It also contains a good presentation of sets. Two sources on discrete structures are Tremblay and Manohar [1975] and Preparata and Yeh [1974].

Aho, Hopcroft, and Ullman [1974] and Knuth [1973, 1974] are excellent references on algorithms and data structures. The topic of undecidability is covered in Hopcroft and Ullman [1969], for example.

Basili and Turner [1975] give a complete description of the programming language SIMPL. Our Pidgin SIMPL is similar in idea to the Pidgin ALGOL in Aho, Hopcroft, and Ullman [1974].

The most classical, complete, and fundamental reference on lattice theory, although advanced, is Birkhoff [1967]. Other references on lattice theory include Szsaz [1963], Gratzer [1971], and Rutherford [1965].

Theorem 2.1 and Algorithm 2.2 can be found in Tenenbaum [1974] and Rosen [1975a], for example, but are not due to these authors.

The observations about properties of operations on semilattices can be found in Kildall [1973], Tenenbaum [1974], Kam and Ullman [1974], Kam and Ullman [1975], Graham and Wegman [1975], and Scott [1970], but again are not due to these authors.

Part II
Control Flow Analysis

Chapter 3

FLOW GRAPHS, DOMINANCE, REDUCIBILITY BY INTERVALS, AND DEPTH−FIRST SPANNING TREES

Our study of control flow analysis is based on a theory of "reducible" flow graphs. That is, in Chapters 3 through 6 we discuss control flow graphs and a structural property of flow graphs called reducibility. Thus, even though control flow analysis touches on both the areas of graph theory and programming languages, our emphasis is more in the realm of graph theory.

Why is reducibility important? Quite simply, the efficiency of many known algorithms for intraprocedural data flow analysis (e.g., Cocke-Allen interval analysis, Ullman's algorithm, and the Graham-Wegman algorithm) is based on flow graphs having this property. Intuitively, reducible flow graphs are very amenable to a divide-and-conquer approach for intraprocedural data flow analysis that is based on various reductive graph transformations.

In this chapter we first define the term "flow graph" and the concept of "dominance". Then, the notion of reducibility by "intervals" is presented. Next we give and discuss an algorithm for partitioning a flow graph into intervals. Finally, depth-first search, a fundamental tool in analyzing graphs, is introduced and the "loop-connectedness" of a flow graph with respect to a depth-first spanning tree of that flow graph is defined.

The concept of dominance is useful in code improvement algorithms for code motion and reduction in strength. Although such algorithms are not included in this book, we do present some algorithms to compute dominators in Chapters 5 and 9. Our primary use of dominance is in analysis and correctness proofs of some data flow analysis algorithms.

The material in this chapter on intervals is essential to the understanding of the interval analysis algorithm in Chapter 8.

Depth-first search is used in three places in this book. In Chapter 4 we use it to help give a theoretical characterization of reducibility. In Chapter 5 we propose rPOSTORDER as a practical node order for the iterative algorithm on problems such as "available expressions". In Chapter 10 we use Tarjan's idea of computing strongly connected components by depth-first search to form transitive closures of the variables that may be modified in each procedure and of certain aliasing information.

The loop-connectedness of a flow graph figures into the analysis of the iterative algorithm on simple problems (Lemmas 7.2 and 7.3, Theorem 7.3). Also, in Section 4.6 we present a theoretically interesting relationship between the loop-connectedness and the "interval derived sequence length".

3.1 Flow Graphs

DEFINITION A *flow graph* is a triple $G=(N,A,s)$, where (N,A) is a (finite) directed graph, and there is a path from the *initial node*, $s \in N$, to every node.

A flow graph is an abstract graph model of a control flow graph with statements, instructions, or quadruples inside each basic block. For the time being (and until Chapter 7) we ignore the possible "contents" of each node and focus our attention on the graph structure of flow graphs.

The definition of flow graph captures two typical and reasonable properties of control flow graphs: there is a specific node at which to begin, and every node is accessible from this initial node. Each of these properties can be guaranteed as follows. If the corresponding procedure has multiple "entries", then a single initial node acting like a "supersource" can be added with an arc to each entry. Any part of the graph inaccessible from the intital node can be removed WLOG because it represents unreachable statements. Note that we allow the in-degree of s to be nonzero. Also, a flow graph need not have any exit nodes. An *exit node of a flow graph* has no successors. Whenever we assume otherwise, we shall explicitly state such assumptions.

Example 3.1 Figure 3.1 contains a flow graph. Node 1 is the initial node, and node 4 is an exit node. We follow the convention of drawing the initial node of a flow graph at the top of its diagram.

An important restriction on flow graphs follows from the nature of branches in programs.

DEFINITION A flow graph in which[1] $r = O(n)$ is called a *sparse flow graph*.

In reality, virtually all flow graphs resulting from programs are sparse because (a) binary branching is generally used for control flow, (b) programmers use disciplined and sparse control flow structures for con-

---

[1]Recall that for a flow graph $G=(N,A,s)$, $n = \#N$ and $r = \#A$.

Dominance

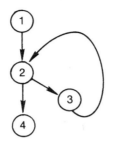

Figure 3.1 A flow graph.

ceptual simplicity, and (c) most programs are written to solve real problems rather than to generate anomalously complicated flow graphs. When no branching more complex than binary is used, $r \leqslant 2n$. Even flow graphs of programs containing case statements are (almost always) sparse.

The significance of sparseness is reflected by algorithmic complexity. Since at worst, $r = O(n^2)$, an algorithm with time complexity $O(r)$ for a flow graph is then $O(n^2)$. If sparseness is assumed, $O(r) = O(n)$.

3.2 Dominance

We shall make pervasive use, although mostly in proofs, of the concept of "dominance" between nodes in a flow graph.

DEFINITION If x and y are two (not necessarily distinct) nodes in a flow graph G, then x *dominates* y iff every path in G from its initial node to y contains x. For convenience, we let $\mathrm{DOM}(y) = \{x \mid x \text{ dominates } y\}$, for each node y. We say that x *properly dominates* y iff $x \neq y$ and x dominates y.

We say that x *directly dominates* (or *immediately dominates*) y iff

(a) x properly dominates y, and
(b) if z properly dominates y, and $z \neq x$, then z (properly) dominates x.

Here are some properties of the dominance relation.

LEMMA 3.1 $\mathrm{DOM}(s) = \{s\}$.

Proof Obvious. □

LEMMA 3.2 *The dominance relation of a flow graph G is a partial ordering.*

Proof

(a) *Reflexivity*. For each node x, it follows from the definition of 'path' that x dominates x.

(b) *Antisymmetry*. Suppose, in contradiction, that dominance is not antisymmetric. Then there exist nodes a and b such that a dominates b, b dominates a, and $a \neq b$. Since $a \neq b$, neither a nor b is s by Lemma 3.1. Let $(w_1, w_2, \ldots, w_k)$ be any cycle-free path from $s = w_1$ to $b = w_k$. One such path exists, since we assume that all nodes are accessible from the initial node. Since a dominates b, there exists an i, $1 < i < k$, such that $a = w_i$. Since b dominates a, there exists a j, $1 < j < i$, such that $b = w_j$. But then b occurs twice on a cycle-free path, a contradiction.

(c) *Transitivity*. Assume that x dominates y, and y dominates z. Let $(w_1, w_2, \ldots, w_k)$ be any path from $s = w_1$ to $z = w_k$. Since y dominates z, there exists a j such that $1 \leq j \leq k$ and $y = w_j$. Since x dominates y, there exists an i such that $1 \leq i \leq j$ and $x = w_i$. But this means that there exists an i such that $1 \leq i \leq k$ and $x = w_i$. That is, x dominates z. □

LEMMA 3.3 *The initial node s of a flow graph G dominates all nodes of G.*

Proof Obvious. □

LEMMA 3.4 *The dominators of a node form a chain (i.e., a linear ordering).*

Proof We shall show that for any three distinct nodes x, y, z, if x and y dominate z, then either x dominates y or conversely.

Let $(w_1, w_2, \ldots, w_k)$ by any cycle-free path from $s = w_1$ to $z = w_k$. By hypothesis, $w_i = x$ and $w_j = y$ for some i and j. WLOG assume $i < j$. We claim that x dominates y.

In proof, suppose that x does not dominate y. Then there is a path P from s to y that does not contain x. Now P concatenated with the path $(w_{j+1}, \ldots, w_k)$ is a path from s to z that does not contain x. But this contradicts the hypothesis that x dominates z. □

LEMMA 3.5 *Every node except s has a unique direct dominator.*

Proof Let x be any node, $x \neq s$. Then $\{s, x\} \subseteq \mathrm{DOM}(x)$. By Lemma 3.4, the dominance relation is a chain on $\mathrm{DOM}(x)$ and also on $\mathrm{DOM}(x) - \{x\}$. Thus, $\mathrm{DOM}(x) - \{x\}$ has a unique least element by Observation 2.5, which must be the direct dominator of x. □

LEMMA 3.6 *The graph of the reflexive and transitive reduction of the dominance relation of a flow graph is a tree. There is an arc (x, y) in the tree if and only if x directly dominates y, and there is a path from u to v in the tree if and only if u dominates v.*

Reducibility by Intervals

Proof By Lemmas 3.1–3.5. □

Example 3.2 In Figure 3.2 we show a flow graph and its dominance tree.

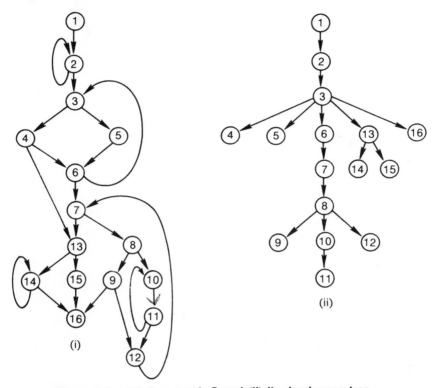

Figure 3.2 (i)A flow graph G and (ii) its dominance tree.

Later on we shall present two dominator algorithms; one for "reducible" flow graphs in Chapter 5 and one for arbitrary flow graphs in Chapter 9.

3.3 Reducibility by Intervals

One approach to intraprocedural data flow analysis is to progressively propagate local information outward from a procedure's innermost "loops" to its outermost "loops", and then reverse the process. Intuitively, this is somewhat analogous to hierarchical bottom-to-top-to-bottom information movement in an organization.

In terms of control flow analysis, the fundamental problem is to define the concept of a "loop". The next problem is to algorithmically identify "loops", construct an innermost to outermost "loop" processing order, and provide a processing order for nodes in a "loop".

First attempts to define a loop as a cycle or a strongly connected component run into trouble because the former gives too fine and the latter too coarse a representation of loops. With cycles, loops are not necessarily properly nested or disjoint; and with strongly connected components, there is no nesting. "Strongly connected regions" are a possible alternative. A *strongly connected region* of a graph G is a subgraph of G in which there is a path between all pairs of nodes. But alas, strongly connected regions do not necessarily cover the graph, nor is a cover necessarily unique. Another alternative for the definition of a loop is an "interval": a construct with many desirable properties.

3.3.1 Intervals

DEFINITION Let G be a flow graph and let h be a node of G. The *interval with header h*, denoted by $I(h)$, is the subset of nodes of G constructed as follows:

$I(h) := \{h\}$ /* initially */
while $\exists$ a node m such that $m \notin I(h) \land m \neq s \land$ all arcs entering m leave nodes in $I(h)$ **do**
$I(h) := I(h) \cup \{m\}$
endwhile

Observe that although node m in the **while**-loop above may not be well defined, $I(h)$ does not depend on the order in which candidates for node m are chosen. A candidate at one iteration of the **while**-loop will, if it is not chosen, still be a candidate at the next iteration.

However, the order in which nodes are added to an interval, called "interval order", is important. If the nodes of an interval are "processed" in interval order, then all interval predecessors of a node are processed before that node.

DEFINITION Let $G=(N,A,s)$ be a flow graph, and let $I(h)$ be an interval of G containing node x. If $(x,h) \in A$, then (x,h) is called a *latching arc* or simply a *latch*.

Example 3.3 $I(2) = \{2,3,4,5\}$ in Figure 3.3. Node 6 cannot be added to $I(2)$ because one of its predecessors, 8, is not in $I(2)$. Arc $(8,6)$ is a latch, for example.

Although we have defined an interval $I(h)$ of a flow graph $G=(N,A,s)$ as a set of nodes, we shall sometimes think of $I(h)$ as the subgraph generated by $I(h)$.

Reducibility by Intervals

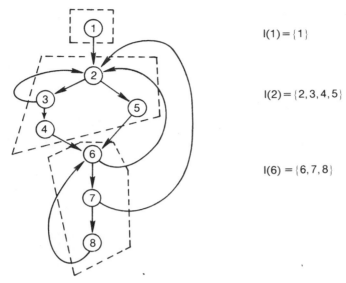

$I(1) = \{1\}$

$I(2) = \{2, 3, 4, 5\}$

$I(6) = \{6, 7, 8\}$

Figure 3.3 Examples of intervals.

DEFINITION A subgraph of a flow graph that is itself a flow graph is called a *subflowgraph*.

LEMMA 3.7 *An interval $I(h)$ of a flow graph $G = (N, A, s)$ is a subflowgraph of G. That is, $\left(I(h), A \cap I(h)^2, h\right)$ is a flow graph.*

Proof By the definition of 'interval', there is a path from h to every node in $I(h)$. □

Since $I(h)$ is a flow graph, h dominates (WRT[2] $I(h)$ alone) all nodes in $I(h)$. However, does h dominate (WRT all of G) all nodes in $I(h)$? We resolve this question and give several important properties of intervals in the next lemma.

LEMMA 3.8 *Let $G = (N, A, s)$ be a flow graph, and let $I(h)$ be an interval of G.*

(a) *Every arc entering a node of the interval $I(h)$ from the outside enters the header h; that is, an interval is single-entry.*
(b) *The header h dominates (WRT all of G) every other node in $I(h)$.*

---

[2] We use 'WRT' as a acronym for 'with respect to'.

(c) *For each node h of a flow graph G, the interval $I(h)$ is unique and independent of the order in which candidates for m in the definition of interval are chosen.*
(d) *Every cycle in an interval $I(h)$ includes the interval header h.*

Proof

(a) Suppose there were another node k in $I(h)$, $k \neq h$, which was also an entry node. That is, there is a node m not in $I(h)$ and there exists an arc (m, k). Thus, k has a predecessor m not in the interval. But this is impossible because k became a member of $I(h)$ only when all of its predecessors were already interval members. (The header h is the only member of $I(h)$ that may have predecessors not in $I(h)$.)

(b) If h does not dominate (WRT all of G) every other node in $I(h)$, there is a node k in $I(h)$ and a path P from the initial node to k that does not contain h. P contains an arc (u, v), where u is not in $I(h)$ and v is. But $v \neq h$ contradicts (a).

(c) Obvious.

(d) Suppose $I(h)$ has a cycle $(x_1, \ldots, x_k)$ that excludes h. Let x_i be the first of $x_1, \ldots, x_k$ added to $I(h)$. Then x_{i-1} (or x_k, if $i=1$) was in $I(h)$ at that time, in contradiction. □

3.3.2 Partitioning a Flow Graph into Intervals

There is a certain way to choose interval headers so that a flow graph is uniquely partitioned into (disjoint) intervals. First we present a readable but undetailed version of this algorithm that we prove to be correct. Then we specify more detail in a second version of the algorithm and show that it can be implemented in $O(r)$ steps.

ALGORITHM 3.1 *Partitioning a flow graph into intervals (version 1).*
INPUT: A flow graph $G = (N, A, s)$, represented by successor lists.
OUTPUT: A set of disjoint intervals whose union is N.
METHOD: See procedure FIND$INTERVALS in Figure 3.4. □

Example 3.4 Shown in Figures 3.3 and 3.5 are the interval partitions of two flow graphs. Intervals are boxed with dashed lines.

THEOREM 3.1 *Algorithm 3.1 constructs a set of disjoint intervals whose union is all the nodes in the flow graph.*

Reducibility by Intervals

procedure FIND$INTERVALS(flow graph $G=(N,A,s)$)
   sets H, /* set of potential header nodes */
       L, /* set of intervals */
   H := {s}
   L := ∅
   while $H \neq \emptyset$ **do**
      Select and delete a node h from H.
      Compute $I(h)$ from the definition of interval.
      Add $I(h)$ to L. /* L is a set of sets. */
      Add to H any node that has a predecessor in $I(h)$, but that is not already in
      H or in one of the intervals of L.
   endwhile
   Output L.
return

Figure 3.4 Algorithm for partitioning a flow graph into intervals (version 1).

Proof If a node is added to an interval, that node subsequently will not be added to the list of potential headers. Thus, disjointness now follows from Lemma 3.8(c).

Since G is a flow graph, every node is accessible from the initial node of G, and so is placed either on the list of potential headers or in an interval. Unless a node is added to an interval, it will become the header of its own interval. So, the union of all intervals is the set of nodes of G. □

To show that FIND$INTERVALS can be implemented in $O(r)$ steps, we now consider a second version of it with more details added.

ALGORITHM 3.2 *Partitioning a flow graph into intervals (version 2).*
   INPUT: A flow graph $G=(N,A,s)$.
   OUTPUT: A set of disjoint intervals whose union is N.
   DATA STRUCTURES: The input flow graph is represented by successor lists. Each interval $I(h)$, represented by a singly linked list, is output "on the fly". The set H of potential headers is represented by a doubly linked list. Four node labels are used; namely, $count[x]$, $reach[x]$, $onH[x]$, and $where[x]$. The $count[x]$ is the number of arcs entering x not yet traversed. The $reach[x]$ eventually becomes the first interval header h such that there is an arc from some node in $I(h)$ to x. If x is on H, then $onH[x]$ is 1; otherwise, $onH[x]$ is 0. If $onH[x]$ is 1, then $where[x]$ points to the particular "node" on H containing x.
   METHOD: See procedure FIND$INTERVALS in Figure 3.6. We have chosen to omit details about $onH[x]$ and $where[x]$. □

THEOREM 3.2 *FIND$INTERVALS, the algorithm to partition a flow graph into intervals, can be implemented in $O(r)$ steps.*

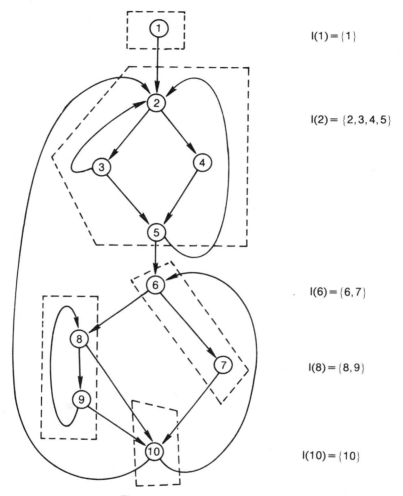

$I(1) = \{1\}$

$I(2) = \{2, 3, 4, 5\}$

$I(6) = \{6, 7\}$

$I(8) = \{8, 9\}$

$I(10) = \{10\}$

Figure 3.5 Interval partition.

Proof Referring to Figure 3.6, initialization may take $O(r)$ steps (and not $O(n)$), if the successor lists are scanned to compute in-degrees. After initialization, each arc is processed exactly once by the next **for**-loop, and a constant number of steps are performed for each arc. This yields an aggregate total of $O(r)$ steps. And $O(r) + O(r)$ is $O(r)$.

To see that a constant number of steps are expended on each arc, note that the following operations cost only a constant number of steps:

(a) selecting and deleting the node from the front of H;
(b) adding a node to the rear of a list;

Reducibility by Intervals

```
procedure FIND$INTERVALS(flow graph G = (N, A, s))
   singly linked list I(h)
   doubly linked list H    /* of potential header nodes */
   integer arrays count [1 : n], reach [1 : n]
   /*** Initialize. ***/
   for each node x do
      count[x] := the in-degree of node x
      reach[x] := 0   /* that is, undefined */
   endfor
   reach[s] := s
   H := {s}   /* WLOG, think of H as a set. */
   /*** Iterate. ***/
   while H ≠ ∅ do
      Select and delete a node h from the front of H.
      /* Now, construct I(h). */
      Initially, I(h) := {h}, and h is "unmarked".
      while ∃ an "unmarked" node x on I(h) do
         Select an "unmarked" node x on I(h) and "mark" x.
         for each arc (x, y) do
            count[y] := count[y] − 1
            if reach[y] = 0 then
               reach[y] := h
               if count[y] = 0 then
                  Add y to the rear of I(h).
               else
                  Add y to the rear of H if it is not already there.
               endif
            else
               if reach[y] = h ∧ count[y] = 0 then
                  Add y to the rear of I(h).
                  Remove y from H if it is there.
               endif
            endif
         endfor
      endwhile
      Output I(h).
   endwhile
   return
```

Figure 3.6 Algorithm for partitioning a flow graph into intervals (version 2).

(c) determining if node y is on H, and if so, where (using onH[y] and where[y]); and

(d) deleting a node y from the doubly linked list H, using where[y].

☐

If the set H of potential header were represented by a bit vector instead of a doubly linked list, then the arrays 'onH' and 'where' are unnecessary. However, this representation for H does not guarantee that operation (a) above takes at most a constant number of steps.

3.3.3 Reducibility

The intervals on one flow graph can be considered as the nodes of another flow graph in which there is an arc between intervals J and K if and only if $J \neq K$, and there is an arc from a node in J to the header of K. Furthermore, this process may be performed repeatedly.

DEFINITION If $G = (N, A, s)$ is a flow graph, then the *derived flow graph* of G, denoted by $I(G)$, is defined as follows:

(a) The nodes of $I(G)$ are the intervals of G.
(b) There is an arc from the node representing interval J to that representing K if there is any arc from a node in J to the header of K, and $J \neq K$.
(c) The initial node of $I(G)$ is $I(s)$.

DEFINITION The sequence $G = G_0, G_1, \ldots, G_k$ is called the *derived sequence* for G iff $G_{i+1} = I(G_i)$ for $0 \leq i < k$, $G_{k-1} \neq G_k$, and $I(G_k) = G_k$. G_k is called the *limit flow graph* of G and is denoted by $\hat{I}(G)$.

DEFINITION A flow graph is called *reducible* (an *RFG*) if and only if its limit flow graph is a single node with no arc (henceforth called the *trivial flow graph*). Otherwise, it is called *irreducible* (an *IRFG*) or *nonreducible*.

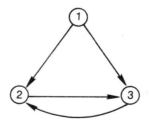

Figure 3.7 The paradigm irreducible flow graph.

Example 3.5 The flow graph in Figure 3.7 is a limit flow graph and is nontrivial. Hence, it is irreducible.

The derived sequence of another flow graph is shown in Figure 3.8. Since its limit flow graph is trivial, this flow graph is reducible.

Reducibility by Intervals

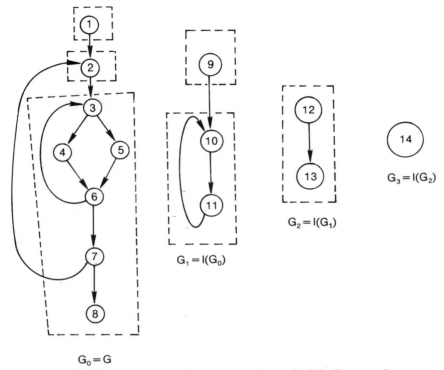

$G_0 = G$

Figure 3.8 The derived sequence of a reducible flow graph.

3.3.4 Interval Order

The intervals of an RFG define a useful processing order on the nodes of an RFG.

DEFINITION *Interval order for an RFG* is defined recursively as follows:

(1) If $I(G)$ is a single node, then an interval order is an order in which nodes may be added to the lone interval G.
(2) If G is reducible and $I(G)$ is not a single node, then an interval order is formed by:
  (a) Find an interval order for $I(G)$.
  (b) In the order of (a), substitute for each node of $I(G)$ the nodes of G that make up the corresponding interval, themselves in interval order assuming each interval is an RFG with the interval header as initial node.

Interval order for an RFG may not be unique.

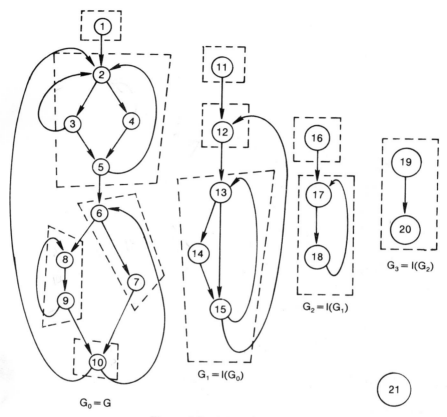

Figure 3.9 Interval order.

Example 3.6 The following sequence illustrates the definition of interval order for the RFG in Figure 3.9.

$$(19, 20)$$
$$(16; 17, 18)$$
$$(11; 12; 13, 14, 15)$$
$$(1; 2, 3, 4, 5; 6, 7; 8, 9; 10)$$

Note that $(1; 2, 4, 3, 5; 6, 7; 8, 9; 10)$ is also an interval order for this RFG.

One property of an interval order of an RFG is that such an order topsorts the dominance relation.

LEMMA 3.9 *Interval order topsorts the dominance relation of an RFG G. That is, if $x \neq y$ and x dominates y in G, then x precedes y in any interval order of G.*

Depth-First Spanning Trees

Proof If x is the initial node s, the lemma is true. Suppose that $x \neq s$. Let $G = G_0, G_1, \ldots, G_k$ be the derived sequence of G.

INDUCTIVE HYPOTHESIS: If x properly dominates y in G_i, then x precedes y in any interval order of the nodes of G_i.

BASIS: $(i = k)$. Vacuously true, since there is only one node in G_k.

INDUCTIVE STEP: $(i < k)$. Assume the inductive hypothesis is true for $i+1$, and consider the case for i. Let x and y be distinct nodes in G_i such that x dominates y. If x and y are in the same interval of G_i, then clearly x precedes y in any interval order of the nodes of G_i.

Now suppose that x and y are in different intervals of G_i; say, intervals J and K respectively.

If J dominates K in G_{i+1}, then J precedes K in any interval order of G_{i+1} by the inductive hypothesis. Consequently, x precedes y in any interval order of G_i.

Suppose that J does not dominate K in G_{i+1}, and $J \neq K$. Let z be the initial node of G_i. Since J does not dominate K, there is a path in G_{i+1} from $I(z)$ to K not containing J. This implies that there is a path in G_i from z to y not containing x, which contradicts the assumption that x dominates y. Thus, this case cannot occur, and this ends the induction.

Finally, applying the inductive hypothesis for $i = 0$ proves the lemma.
□

In summary, intervals have accomplished for us what we originally set out to accomplish in Section 3.3. The derived sequence of an RFG defines a suitable innermost to outermost "loop" processing order; namely, interval order. The use of this for intraprocedural data flow analysis is explained in Chapter 8.

3.4 Depth-First Spanning Trees

"Depth-first spanning trees" are used in this study to help characterize reducible flow graphs in Chapter 4, and also to establish one possible node processing order for an iterative approach to intraprocedural data flow analysis (cf. Chapters 5 and 7). We also employ them in some proofs.

"Depth-first search" has other possible uses for flow analysis. For example, the SCCs of a call graph can be determined in $O(r)$ steps by depth-first search. If a bit vector of information is associated with each procedure in the call graph, say the globals that may be directly modified by each procedure, then the transitive closure of such bit vectors can be performed as a by-product of the SCC routine. (See Algorithm 10.1 in Section 10.4.4, for example.)

DEFINITION A *depth-first spanning tree* (*DFST*) of a flow graph G is a (directed, rooted,) ordered spanning tree grown by Algorithm 3.3 given below. This algorithm also defines an ordering on the nodes of G that we call *rPOSTORDER* (*i.e.*, reverse POSTORDER).

ALGORITHM 3.3 *rPOSTORDER computation by depth-first search.*
  INPUT: A flow graph $G=(N,A,s)$, represented by a successor lists. Initially, the nodes of G are numbered arbitrarily from 1 to n.
  OUTPUT: A numbering of the nodes from 1 to n, in array rPOST-ORDER, indicating the reverse of the order in which each node was last visitited during a depth-first search of G.
  METHOD:
  Initially all nodes are marked "unvisited" in a global array.
  There is a global integer array rPOSTORDER$[1:n]$.
  There is a global integer i with initial value n.
  The algorithm consists of a call to DFS(s), where DFS is the recursive procedure defined in Figure 3.10. □

---

    recursive procedure DFS(**integer** x)
        Mark x "visited".
        while SUC$(x) \neq \emptyset$ **do**
            Select and delete a node y from SUC(x).
            if y is marked "unvisited" **then**
                Add arc (x,y) to the DFST diagram.
                call DFS(y)
            endif
        endwhile
        rPOSTORDER$[x]$:= i
        i := $i-1$
    return

Figure 3.10 rPOSTORDER Algorithm.

DEFINITION If (u,v) is an arc in a DFST, then u is the *parent* of v, and v is a *child* of u. The *ancestor* and *descendant* relations are the respective transitive closures of the parent and child relations.

Let $G=(N,A,s)$ be a flow graph and let $T=(N,A')$ be a DFST of G. The arcs in $A - A'$ fall into three categories:

(a) Arcs that go from ancestors to descendants we call *forward arcs WRT T*.
(b) Arcs that go from descendants to ancestors or from a node to itself we call *back arcs*[3] *WRT T*.
(c) Arcs that go between nodes that are unrelated by the ancestor-descendant relation we call *cross arcs WRT T*.

---

[3]Our 'back arcs' are what Tarjan [1972] calls 'fronds'.

Depth-First Spanning Trees

We follow the convention of drawing trees with the root on the top. We show arcs in T by solid lines directed downward, and arcs of G not in T by dashed lines. The children of a node in T are always ordered from left to right in a diagram.

Note that there may be more than one DFST for a given flow graph G. Thus, the classification of arcs of G into the different DFST arc categories is relative to a fixed DFST of G.

Example 3.7 Let G be the flow graph of Figure 3.11(i). If we consider the nodes in the order 1,2,3,4, then back to 3, then go to 5, we obtain the DFST of Figure 3.11(ii). Arc (1,3) is a forward arc. Arcs (3,2), (4,2), and (5,5) are back arcs. Arc (5,4) is a cross arc.

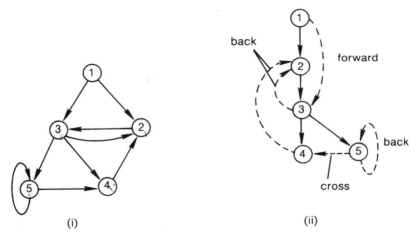

Figure 3.11 Example of a DFST of a flow graph.

Since a DFST is an ordered tree, the children of each node z are ordered from left-to-right so that "younger" children of z are to the right of "older" children of z. We extend the notion of "to the right" in a DFST by saying that if x is to the right of y, then all of x's descendants are to the right of all of y's descendants. Thus, if (u,v) is a cross arc WRT a DFST, then u is to the right of v.

OBSERVATION 3.1 *Let $G=(N,A,s)$ be a flow graph, and let T be a DFST of G.*

(a) *Arc (x,y) is a back arc WRT T iff $rPOSTORDER[x] \geqslant rPOSTORDER[y]$.*
(b) *Every cycle of G contains at least one back arc.*

DEFINITION Let $G=(N,A,s)$ be a flow graph, and let T be a DFST of G. We define $d(G,T)$, the *loop-connectedness* of G WRT T, to be the largest number of back arcs (WRT T) found in any cycle-free path of G.

Exercises

3.1 What features of ALGOL 60 and PL/I complicate the construction of control flow graphs?

3.2 Why are flow graphs usually constructed from an intermediate representation of a program rather than from the source program or its machine language representation?

3.3 Design an efficient algorithm to construct the dominance tree of a reducible flow graph.

3.4 Find a real computer program with a procedure such that the interval derived sequence length of the flow graph of that procedure is at least four.

Bibliographic Notes

Two fundamental references on control flow analysis and dominance are Prosser [1959], and Lowry and Medlock [1969]. Our treatment of dominance also draws from Allen [1970] and Aho and Ullman [1973].

The strongly connected region approach to flow analysis is described in Allen [1969].

Interval analysis was developed by Cocke [1970], and further described in Allen [1970], Cocke and Schwartz [1970], Kennedy [1971], Allen [1971], Allen and Cocke [1976], and Kennedy [1974]. Much of our treatment of intervals comes from Allen [1970]. Algorithm 3.2 is taken from Aho and Ullman [1973].

Material on depth-first search can be found in Tarjan [1972], Hopcroft and Tarjan [1973], and Aho, Hopcroft, and Ullman [1974].

Fong and Ullman [1976] given an $O(n^2)$ algorithm to find the loop-connectedness of a sparse RFG.

For references containing dominator algorithms, see the Bibliographic Notes at the end of Chapter 5.

Chapter 4
CHARACTERIZATIONS OF REDUCIBLE FLOW GRAPHS

It has been experimentally observed by Allen [1970] and Knuth [1971] that virtually all flow graphs of procedures of computer programs occuring "in nature" are reducible.[1] Also, the class of flow graphs of "structured programs" (see Kosaraju [1974], for example) are reducible. Thus, not only is the class of reducible flow graphs a natural class of graphs to study, but it is reasonable to design and use intraprocedural data flow analysis algorithms that are based on the underlying control flow graph being reducible. In fact, Cocke-Allen interval analysis, Ullman's algorithm, and the Graham-Wegman algorithm are based on reducible flow graphs.

In this chapter we present some characterizations of reducible flow graphs. The purpose of this is to gain a better understanding of reducibility, and to see that it is not necessarily tied to intervals. Reducibility is characterized by two very simple reductive graph transformations, a forbidden subgraph, the uniqueness of a certain DAG, and single-entry conditions.

This entire chapter is mainly of theoretical interest and is of no immediate practical use (as evidenced by the absence of algorithms). However, this material is not completely useless! For example, reductive graph transformations T1 and T2 of Section 4.1 form part of the basis for both Ullman's algorithm (also see Fong, Kam, and Ullman [1975]) and the Graham-Wegman algorithm. Certainly, the fact that an idea helps (or may help) in designing a practional algorithm is good justification for its presentation. The concept of a "region" is just another model of a "loop". The concept of a "parse of a reducible flow graph" is important because it suggests that algorithms for flow analysis using "graph-transformation directed translations" (analogous to the "syntax directed translations" familiar to context-free language theory) are possible. Also, some of the results in this chapter help simplify the analysis of some intraprocedural data flow analysis algorithms.

---

[1] It is interesting to note that Knuth [1974] contains several programs with irreducible flow graphs. Can you find them?

4.1 Graph Transformations T1 and T2

Although reducibility has been defined in terms of intervals (in Chapter 3) and the notion of an RFG parse seems to be exclusively associated with intervals, this is not the case. In this section we shall introduce two graph transformations, called T1 and T2, and show that reducibility by T1 and T2 (which we temporarily call "collapsibility") is equivalent to reducibility by intervals.

4.1.1 Collapsibility

DEFINITION Let G be a flow graph and let (w,w) be an arc of G. *Transformation T1* is the removal of this arc. Formally, if $G=(N,A,s)$ and $(w,w) \in A$, then $T1(G,(w,w)) = G'$, where $G' = (N, A - \{(w,w)\}, s)$. We also indicate this by writing $G \underset{T1}{\Rightarrow} G'$.

DEFINITION Let G be a flow graph, and let y not be the initial node[2] and have a single predecessor, x. *Transformation T2* is the replacement of x, y and (x,y) by a single node z. Predecessors of x become predecessors of z. Successors of x or y become successors of z. There is an arc (z,z) if and only if there was formerly an arc (y,x) or (x,x). (Whenever T2 is applied as described here, we say that x consumes y.)

Formally, if

$$G = (N, A, s),$$
$$x, y \in N,$$
$$y \neq s,$$
$$(x,y) \in A,$$
$$\neg (\exists w)[(w,y) \in A \wedge w \neq x],$$

then

$$T2(G,(x,y)) = G' \quad \left(\text{also } G \underset{T2}{\Rightarrow} G'\right),$$

where

$$G' = (N', A', t),$$
$$z \notin N$$
$$N' = (N - \{x, y\}) \cup \{z\}.$$

Let $f: N \rightarrow N'$ be defined by $f(x) = f(y) = z$, and $f(v) = v$ for $v \neq x, y$.

[2] The stipulation that $y \neq s$ is necessary in the definition of T2 because we allow arcs to enter s.

Then

$$t = f(s),$$

$$A' = \{(f(u), f(v)) | (u,v) \in A \wedge (u,v) \neq (x,y)\}.$$

Example 4.1 Figure 4.1 shows a flow graph that is transformed into the trivial flow graph by one application of T1 followed by two applications of T2. Although T2 is not applicable to the original flow graph, it becomes applicable after use of T1.

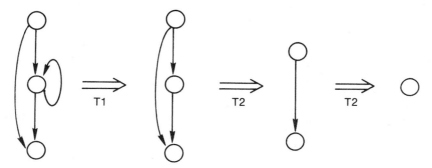

Figure 4.1 Applications for T1 and T2.

DEFINITION A flow graph is called *collapsible* if and only if it can be transformed into the trivial graph by repeated application of T1 and T2. Otherwise, it is called *non-collapsible*.

Example 4.2 The flow graph of Figure 4.1 is collapsible, whereas the one in Figure 3.7 is not.

We shall now prove that T1 and T2 together form a "finite Church-Rosser" transformation. This means that they may only be applied a finite number of times to any flow graph, and the final result is independent of the order in which they were applied. The "order independence" (or "commutativity") part of this property is important for two reasons. First and most important, it allows flexibility in algorithm design. Second, it greatly simplifies proofs.

DEFINITION A pair $(S, \Rightarrow)$, where S is a set and $\Rightarrow$ is a relation on S, is said to be *finite* iff for each p in S, there is a constant k_p such that if $p \stackrel{i}{\Rightarrow} q$, then $i \leq k_p$. That is, there is a bound on the number of times $\Rightarrow$ can be applied in succession, beginning with any element p. We say that $(S, \Rightarrow)$ is

Graph Transformations T1 and T2

finite Church-Rosser (acronym: *FCR*) if it is finite, and[3] $\hat{\Rightarrow}$ is a function; i.e., $p \hat{\Rightarrow} q$ and $p \hat{\Rightarrow} r$ implies $q = r$. If set S is understood, $\Rightarrow$ is called an *FCR transformation*.

The following result (proved elsewhere and presented as a "test") gives a test for the FCR property that is easier to apply than the above definition.

Test for FCR property *If $\Rightarrow$ is a relation on a set S, then $(S, \Rightarrow)$ is FCR if and only if it is finite, and for all p in S, if $p \Rightarrow q$ and $p \Rightarrow r$, then there is some t such that $q \stackrel{*}{\Rightarrow} t$ and $r \stackrel{*}{\Rightarrow} t$.*

DEFINITION Let S be the set of flow graphs. We define the relation $\underset{Ti}{\Rightarrow}$, $i \in \{1,2\}$, by $G \underset{Ti}{\Rightarrow} G'$ if and only if G can be transformed into G' by one application of Ti. Let $\Rightarrow$ denote the union of $\underset{T1}{\Rightarrow}$ and $\underset{T2}{\Rightarrow}$. The reflexive closure, k-fold product, transitive closure, reflexive transitive closure, and the completion of $\Rightarrow$ are respectively given by $\stackrel{\#}{\Rightarrow}$, $\stackrel{k}{\Rightarrow}$, $\stackrel{+}{\Rightarrow}$, $\stackrel{*}{\Rightarrow}$ and $\hat{\Rightarrow}$.

THEOREM 4.1 *$(S, \Rightarrow)$ is FCR.*

Proof We use the above test and note that in this case, we shall always be able to find t such that $q \stackrel{\#}{\Rightarrow} t$ and $r \stackrel{\#}{\Rightarrow} t$.

FINITENESS PROPERTY: Let G be a flow graph with n nodes. Each application of T1 or T2 deletes at least one arc. Thus, $\Rightarrow$ is finite.

COMMUTATIVITY PROPERTY: Suppose that $G \underset{Ti}{\Rightarrow} L$ and $G \underset{Tj}{\Rightarrow} M$, where $i, j \in \{1,2\}$. Then the situation depicted in Figure 4.2 always holds. There are three distinct cases to consider.

CASE 1: $i = j = 1$. Suppose that T1 is applied to node w to yield L and to node x to yield M. If $w = x$, then $L = M$. If $w \neq x$, then T1 may be performed on x in L and on w in M to yield equal graphs.

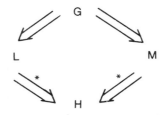

Figure 4.2 "Commutativity" of $\Rightarrow$.

---

[3]Recall, from Section 2.1.4, that '^' above a relation denotes the completion of that relation.

Thus, $G \Rightarrow L \stackrel{\#}{\Rightarrow} H$ and $G \Rightarrow M \stackrel{\#}{\Rightarrow} H$, where H is the graph resulting after applying T1 to nodes w and x in G.

CASE 2: $i=j=2$. Suppose that T2 is applied to nodes w and x in G, with w consuming x, to yield L, and to nodes y and z in G, with y consuming z, to yield M. If $w=y$ and $x=z$, then $L=M$. If all four nodes are distinct, then let y consume z in L, and let w consume x in M to yield equal graphs. Now suppose that neither of the previous subcases holds. If $w=y$ and no other equalities hold, then Figure 4.3 explains this. Otherwise, if $x=y$ and no other equalities hold, then Figure 4.4 explains this situation. Thus, $G \Rightarrow L \stackrel{\#}{\Rightarrow} H$ and $G \Rightarrow M \stackrel{\#}{\Rightarrow} H$, where H is the graph resulting from w consuming x and y consuming z in G.

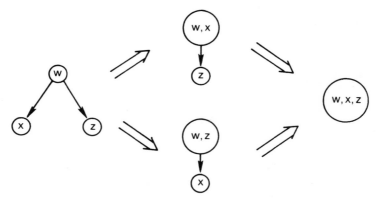

Figure 4.3 Applications of T2.

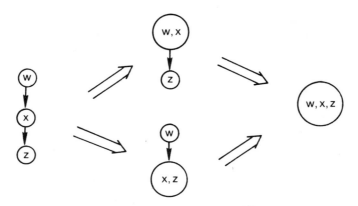

Figure 4.4 Applications of T2.

Graph Transformations T1 and T2

The case in which $w = z$ and no other equalities hold is symmetric to the case $x = y$ above. The case $w = z$ and $x = y$ is impossible, because then the flow graph has two nodes isolated from the others, and hence must consist of only w and x. But, one of these must be the initial node, and thus either w cannot consume x, or y cannot consume z. Since we have assumed $w \neq x$ and $y \neq z$, and x may not be z unless $w = y$, we have considered all possibilities.

CASE 3: $i \neq j$. Suppose that T2 is applied so that w consumes x in G to yield L, and T1 is applied to node z in G to yield M. Clearly, $y \neq z$. Consequently, T1 and T2 do not "interfere"; T1 may be applied to node z in L, and T2 may be applied to nodes x and y (with x consuming y) in M to yield equal graphs. Thus, $G \Rightarrow L \stackrel{\#}{\Rightarrow} H$ and $G \Rightarrow M \stackrel{\#}{\Rightarrow} H$, where H is the result of x consuming y and applying T1 to z in G. □

4.1.2 Equivalence of Reducibility and Collapsibility

In this section we show that a flow graph is reducible if and only if it is collapsible.

DEFINITION For each $k \geq 1$, let the first k nodes in an interval order of $I(h)$ be called a *partial interval*.

LEMMA 4.1 *If G is a flow graph, then $G \stackrel{*}{\Rightarrow} I(G)$.*

Proof If suffices to show that a partial interval is collapsible to its header, and that connections (arcs) between a partial interval and the other nodes in the flow graph are maintained. Thus, constructing the derived graph $I(G)$ of a flow graph G corresponds exactly to collapsing each interval of G.

INDUCTIVE HYPOTHESIS: A partial interval of k nodes is collapsible to its header, and arcs between the partial interval and the other nodes of the flow graph are preserved. That is, arcs leaving the partial interval to a node outside the partial interval remain. The header has no self-loop.

BASIS: The first node added to an interval is the header node. The only collapsing possible is removal of a self-loop if present. This possible application of T1 will not destroy any arc to another node in the graph outside the partial interval.

INDUCTIVE STEP: Assume that the inductive hypothesis is true for a partial interval of k nodes, and consider the addition of another node x to the partial interval. This new node only has arcs entering it from nodes in the partial interval. Since the first k nodes of the partial interval are collapsible by the inductive hypothesis, there will be exactly one arc from the collapsed partial interval to x. Moreover, x cannot be the initial node.

Thus, T2 is applicable. Arcs from x to nodes outside the partial interval now leave the node representing the collapsed partial interval. If there is a self-loop introduced by the application of T2, it can be removed by T1. □

As an immediate consequence of Lemma 4.1, we have the following.

THEOREM 4.2 *If a flow graph is reducible, then it is collapsible.*

Proof If $\hat{I}(G) = 0$,[4] then $G \stackrel{*}{\Rightarrow} 0$ by Lemma 4.1, iterated. □

The converse of Theorem 4.2 is easy to prove.

THEOREM 4.3 *If a flow graph is collapsible, then it is reducible.*

Proof Suppose $G \stackrel{*}{\Rightarrow} 0$, and let $\hat{I}(G) = G'$. By Lemma 4.1 iterated, $G \stackrel{*}{\Rightarrow} G'$. We must have $G' \stackrel{*}{\Rightarrow} 0$. (For if $G' \stackrel{*}{\Rightarrow} G''$, then $G \stackrel{*}{\Rightarrow} G''$. Since $\stackrel{*}{\Rightarrow}$ is a function, and $G \stackrel{*}{\Rightarrow} 0$, we have $G'' = 0$.)

If $G' \neq 0$, then since $G' \stackrel{*}{\Rightarrow} 0$, T1 or T2 is applicable to G'. We have assumed $I(G') = G'$, so every node appears on the header list when Algorithm 3.1 is applied to G'. If T1 is applicable to node x, then the node representing $I(x)$ does not have a self-loop in $I(G')$. Therefore, $I(G') \neq G'$. If T2 is applicable to nodes x and y, with x consuming y, then y is in $I(x)$ and x appears on the header list before y, so again, $I(G') \neq G'$. We conclude that $G' = 0$. □

DEFINITION A graph property, say P, is said to be *preserved* by a graph transformation iff, whenever the graph prior to the transformation has property P, then the graph after the transformation has property P.

COROLLARY 4.1 *T1 and T2 preserve both reducibility and irreducibility.*

Proof The result is clearly true for T1. If T2 did not preserve reducibilty (and irreducibility), then $\stackrel{*}{\Rightarrow}$ would not be a function, a violation of Theorem 4.1. To see this, an argument similar to the first part of the proof of Theorem 4.3 suffices. □

4.2 A Forbidden Subgraph for Reducible Flow Graphs

We now show the existence of a certain family of subgraphs in all and only the irreducible flow graphs.

DEFINITION Let (*) denote any of the flow graphs represented in Figure 4.5, where the wavy lines denote arc disjoint paths; nodes a, b, c, and s, the initial node, are distinct, except that a and s may be the same.

---
[4]Let the symbol '0' denote the trivial flow graph.

A Forbidden Subgraph for Reducible Flow Graphs

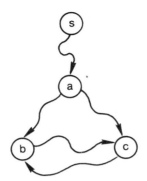

Figure 4.5 The (*) subgraph.

LEMMA 4.2 *The absence of subgraph (*) in a flow graph is preserved by T1 and T2.*

Proof Let G be a flow graph, and let x and y be two nodes in G. We observe that if a path does not exist between x and y, then neither T1 nor T2 will create such a path; neither will they make two paths be arc disjoint if they were not so already. □

THEOREM 4.4 *If a flow graph is irreducible, then it contains a (*) subgraph.*

Proof We prove the theorem by induction on n, the number of nodes in the flow graph.
  INDUCTIVE HYPOTHESIS: If a flow G with n nodes is irreducible, then it contains a (*) subgraph.
  BASIS: ($n=2$). This case is vacuous because an irreducible flow graph must have at least three nodes.
  INDUCTIVE STEP: ($n \geqslant 3$). Assume that the inductive hypothesis is true for all flow graphs up to $n-1$ nodes, and consider an irreducible flow graph G with n nodes.
  By Lemma 4.2, we may assume, without loss of generality, that neither T1 nor T2 is applicable to G. That is, if G can become G' under repeated application of T1 or T2, and we can show that G' has (*), then we will also have shown that G has (*). Thus, we may assume that G is a limit flow graph, and G has at least three nodes.
  Let T be any DFST of G, and let y be the right-most child of the root of T. Node y has at least two entering arcs in G, for otherwise T2 would apply. One of these arcs leaves the root and is an arc in T. All other arcs entering y must be back arcs. They cannot be forward arcs because forward arcs run from ancestors to descendants, the root is the only

ancestor of y, and there already is such an arc. They cannot be cross arcs because cross arcs run from right to left in T, and nothing is to the right of y.

Let (x,y) be a back arc entering y.
Let $L = (N', A')$ be the subgraph of G such that:

(a) N' contains x, y, and all nodes z such that there is a path from z to x that does not pass through y.
(b) $A' = A \cap (N' \times N')$.

There are two cases to consider.

CASE 1: The initial node is in N'. Thus, y does not dominate x. See Figure 4.6(i). There is a forward arc from the initial node to w, where $w = x$ or w is an ancestor of x and $w \neq y$. Because (x,y) is a back arc, y is an ancestor of x and there is a path in T from y to x. If $w = x$, then nodes s, x, y play the roles of a, b, c in (*). If $w \neq x$, then s, w, y form (*). The paths are easily seen to be arc disjoint.

CASE 2: The initial node is not in N'. Thus, y dominates x. See Figure 4.6(ii). We claim that $H = (N', A', y)$ is a flow graph, because y dominates each node z in N'. Furthermore, all arcs entering H from outside enter y.

Thus, any reduction by T1 or T2 taking place in the subflowgraph H, with y treated as the initial node, will also be a valid reduction in G. Since

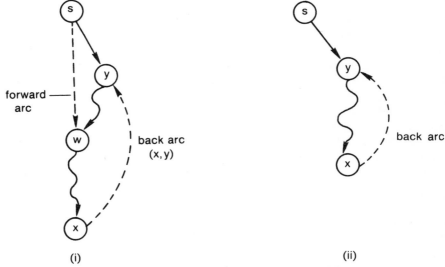

Figure 4.6 Pictures for proof of Theorem 4.4. In (ii), node y is the right-most child of s.

A Forbidden Subgraph for Reducible Flow Graphs

G is a limit flow graph, we conclude that H is also a limit flow graph. Finally, since s is not in H, H has at most $n-1$ nodes and the inductive hypothesis applies. This ends the induction.

But, since H has a (*) subgraph with initial node y, it is easy to show that G has a (*) subgraph with initial node s by adding the path from s to y. □

COROLLARY 4.2 *If G is a nontrivial limit flow graph, then it has a (*) subgraph in which $s = a$, and the paths from a to b and from a to c are each a single arc.*

THEOREM 4.5 *If a flow graph contains a (*) subgraph, then it is irreducible.*

Proof We proceed by induction on the number of nodes, n, in G.

BASIS: ($n = 3$). This is an elementary consideration of the three cases in Figure 4.7 with the initial nodes at the top.

INDUCTIVE STEP: ($n > 3$). Assume the inductive hypothesis for $n-1$ and consider an RFG G with $n > 3$ nodes that has a (*) subgraph. Let G' be the graph formed by applying T1 to G until no longer possible. It is easy to see that G' also contains (*), and is reducible since T1 and T2 are FCR. Therefore T2 is applicable to nodes x and y of G', where x can consume y. Let G'' be the result of x consuming y in G'. We consider cases, depending on the relation of y to (*).

CASE 1: Node y is not one of the nodes represented by (*), including the paths shown. It is straightforward in this case to show that (*) is present in G''.

CASE 2: Node y is a of (*). See Figure 4.5. Then x must be the predecessor of a on the path from the initial node to a. Again, (*) exists in G''.

CASE 3: Node y is b or c. Since b and c each have at least two distinct predecessors, this case is impossible.

CASE 4: Node y is a node of one of the paths of (*). Then x is on the same path (possibly an endpoint). Once more, (*) clearly exists in G''.

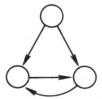 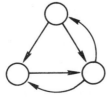 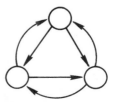

Figure 4.7 Irreducible flow graphs with three nodes.

Since G'' has one fewer node than G, the inductive hypothesis applies to G''. Therefore G'' is irreducible. Since T1 and T2 are FCR, $G \stackrel{*}{\Rightarrow} G''$, and $G \stackrel{*}{\Rightarrow} 0$, it follows that $G'' \stackrel{*}{\Rightarrow} 0$. That is, G'' is reducible. We have a contradiction, and conclude that G is irreducible. □

4.3 Regions, Parses, and Backward Arcs

In this section we prove that if G is an RFG, then arc (x,y) is "backward" if and only if y dominates x in G. To accomplish this, we define concepts for T1 and T2 that are very similar to those for intervals. These concepts include "regions", "parses", and "backward arcs".

DEFINITION Let $G = (N, A, s)$ be a flow graph, let $N_1 \subseteq N$, let $A_1 \subseteq A$, and let h be in N_1. $R = (N_1, A_1, h)$ is called a *region* of G with *header* h iff in every path $(x_1, \ldots, x_k)$, where $x_1 = s$ and x_k is in N_1, there is some $i \leq k$ such that

(a) $x_i = h$; and
(b) $x_{i+1}, \ldots, x_k$ are in N_1; and
(c) $(x_i, x_{i+1}), (x_{i+1}, x_{i+2}), \ldots, (x_{k-1}, x_k)$ are in A_1.

That is, access to every node in the region is through the header only.

Example 4.3 The subgraph boxed with dashed lines in Figure 4.8(i) is a region. The subgraph enclosed by the curved lines in Figure 4.8(ii) is not a region because the path $(1,2,4,5,4)$ violates conditions (b) and (c) in the definition of 'region'.

LEMMA 4.3 *A region of a flow graph is a subflowgraph.*

Proof Obvious. □

LEMMA 4.4 *Let G be a flow graph. The header of a region of G dominates (WRT all of G) all nodes in the region.*

Proof By the definition of 'region'. □

As we proceed to apply T1 and T2 to a flow graph, each arc of an intermediate graph represents a set of arcs, and each node represents a set of nodes and arcs in a natural way.

DEFINITION We say that each node and arc in the original flow graph *represents* itself. If T1 is applied to node w with arc (w,w), then the resulting node *represents* what node w and arc (w,w) represented. If T1 is

Regions, Parses, and Backward Arcs

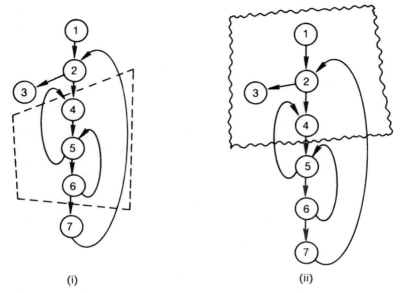

Figure 4.8 Example of a region and a nonregion.

applied to x and y with arc (x,y) eliminated, then the resulting node z *represents* what x, y and (x,y) represented. In addition, if two arcs (x,u) and (y,u) are replaced by a single arc (z,u), then (z,u) *represents* what (x,u) and (y,u) represented.

Example 4.4 Each node and arc in the flow graph of Figure 4.9(i) is uniquely labeled and represents itself. In Figure 4.9(ii), node 4 represents $\{2,b\}$, and arc e represents $\{c,d\}$. Node 6 in Figure 4.9(iv) represents $\{1,2,3,a,b,c,d\}$.

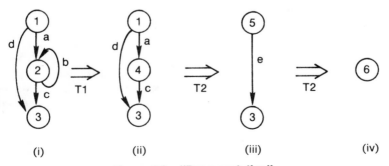

Figure 4.9 "Representation".

The two lemmas that follow establish that a node at some intermediate stage of the reduction of a flow graph is, in fact, a region.

LEMMA 4.5

(a) Let z be a node constructed during the reduction of some flow graph G. If z represents arc (x,y) of G, then x and y are represented by z.

(b) Let w and x be (not necessarily distinct) nodes constructed during the reduction of G and e the arc constructed from w to x. If e represents arc (y,z) of G, then y is represented by w and z by x.

(c) In any graph formed while reducing G, all nodes and arcs represent disjoint sets of objects (nodes and arcs).

Proof Each observation is a straightforward induction on the number of arcs represented by a node or arc. □

LEMMA 4.6 Let $G = (N, A, s)$ be an RFG, and let $N_1 \subseteq N$ and $A_1 \subseteq A$ be a set of nodes and arcs represented by a single node at some stage of the reduction of G. Then there is a node h in N_1 such that (N_1, A_1, h) is a region of G with header h.

Proof We perform induction on the number of arcs in A_1. The basis of zero arcs is easy, since every node forms a region by itself. For the induction step, suppose that the node representing N_1 and A_1 was formed by an application of T1. Then by the inductive hypothesis, there is a proper subset of A_1, say A_2, and a node h in N_1 such that (N_1, A_2, h) is a region. Choose A_2 to be all the arcs of A_1 except those represented by the arc that is eliminated by the final application of T1. Then (N_1, A_1, h) is also a region, since the arcs of $A_1 - A_2$ are all between nodes of N_1. The addition of such arcs cannot cause a violation of the definition of 'region'.

Now suppose the node representing N_1 and A_1 was formed by T2. That is, this node was formed from a node representing N_2 and A_2, a node representing N_3 and A_3 and an arc (from the former to the latter node) representing A_4, where $N_1 = N_2 \cup N_3$ and $A_1 = A_2 \cup A_3 \cup A_4$.

Returning to our argument, we know by the inductive hypothesis that there are nodes u in N_2 and v in N_3 such that (N_2, A_2, u) and (N_3, A_3, v) are regions. It is claimed that (N_1, A_1, u) must also be a region. If not, there must exist a node $w \neq u$ in N_1 and a path $x_1, \ldots, x_k, w, u_1, \ldots, y_r$ such that $x_1 = s$, and either x_k is not in N_1 or (x_k, w) is not in A_1.

CASE 1: x_k is not in N_1. If w is in N_2, we contradict the hypothesis that (N_2, A_2, u) is a region. If w is in N_3, and $w \neq v$, then (N_3, A_3, v) would not be a region. Suppose that $w = v$. Since T2 was applicable to the nodes representing N_2 and N_3, it must be by Lemma 3.5(b) that A_4 includes all

Regions, Parses, and Backward Arcs

arcs of the original graph between N_2 and N_3, and that no other arcs enter nodes in N_3. Thus, x_k is in N_2 and hence in N_1.

CASE 2: (x_k, w) is not in A_1. If w is in N_2, there is a contradiction of the fact that (N_2, A_2, u) is a region. If w is in N_3 and $w \neq v$, we again contradict the fact that (N_3, A_3, v) is a region. If $w = v$, then as we argued above, (x_k, w) must be in A_4, hence in A_1. □

LEMMA 4.7 *In a flow graph, if region R results from region R' consuming region R'', then the header h of R' dominates all nodes in R''.*

Proof It is easy to show that h must be the header of R. Thus, h dominates all nodes in R by Lemma 4.4. □

Since T1 and T2 may be applied to an RFG in different sequences, it becomes necessary (and useful) to discuss certain specific sequences of applications of T1 and T2. Informally, a "parse" of an RFG is a list of the reductions made (T1 or T2) and the regions to which they apply.

DEFINITION A *parse* π of an RFG $G = (N, A, s)$ is a sequence of objects of the form $(T1, u, v, S)$ or $(T2, u, v, w, S)$, where u, v, and w are names of nodes and S is a set of arcs. We define the parse of an RFG recursively as follows:

(a) The trivial flow graph has only the empty sequence as its parse.

(b) If G' (which may not be the original flow graph in a sequence of reductions) is reduced to G'' by an application of T1 to node u, and the resulting node is named v in G'', then $(T1, u, v, S)$ followed by a parse of G'' is a parse of G', where S is the set of arcs represented by the arc (u, u) eliminated from G'.

(c) If G' is reduced to G'' by an application of T2 to nodes u and v (with u consuming v), and the resulting node is called w, then $(T2, u, v, w, S)$ followed by a parse of G'' is a parse of G', where S is the set of arcs represented by the arc (u, v) in G'.

(d) In both (b) and (c) above, "representation" in G' carries over to G''. That is, whatever an object represents in G' is also represented by that object in G'', except for those changes in representation caused by the particular transformation (T1 or T2) currently being applied.

Example 4.5 The parse of the flow graph in Figure 4.9 is as follows:

$$(T1, 2, 4, \{b\})$$
$$(T2, 1, 4, 5, \{a\})$$
$$(T2, 5, 3, 6, \{c, d\}).$$

Although the parse of this flow graph is unique (up to renaming of nodes), not all flow graphs have unique parses.

A given RFG may have many parses, but there are certain things in common among all these parses.

DEFINITION Let $G=(N,A,s)$ be an RFG and let π be a parse of G. We say that an arc in A is a *backward arc WRT* π if it appears in set S of an entry $(T1,u,v,S)$ of π and a *forward arc*[5] *WRT* π otherwise. Let $B(G)$ be the set of arcs in A that are backward in every parse of G.

Example 4.6 Consider Figure 4.10, one of whose parses is as follows:

$$(T2, 2, 3, 8, \{b\})$$
$$(T2, 6, 7, 9, \{h\})$$
$$(T2, 5, 9, 10, \{f\})$$
$$(T1, 10, 11, \{g\})$$
$$(T2, 4, 11, 12, \{d\})$$
$$(T1, 12, 13, \{e\})$$
$$(T2, 8, 13, 14, \{c\})$$
$$(T1, 14, 15, \{i\})$$
$$(T2, 1, 15, 16, \{a\}).$$

The forward arcs of this parse are $\{b,h,f,d,c,a\}$, and the backward arcs are $\{g,e,i\}$.

We shall show that $B(G)$ is the (unique) set of backward arcs of an RFG G, but for the present we shall assume that the backward arcs of two distinct parses of an RFG are not necessarily the same.

LEMMA 4.8 *If $G=(N,A,s)$ is an RFG, then $\{(x,x)\in A\} \subseteq B(G)$.*

Proof Obvious. □

LEMMA 4.9 *If (s,y) is backward, then $y=s$.*

Proof If y is not the initial node, then there exists a region that includes the initial node but does not have it as its header in violation of Lemma 4.7. □

---

[5] Do not confuse this definition of 'forward' arc of an RFG with the previous one for an arc in a DFST. They are not necessarily the same, and context should distinguish which one is meant.

Regions, Parses, and Backward Arcs

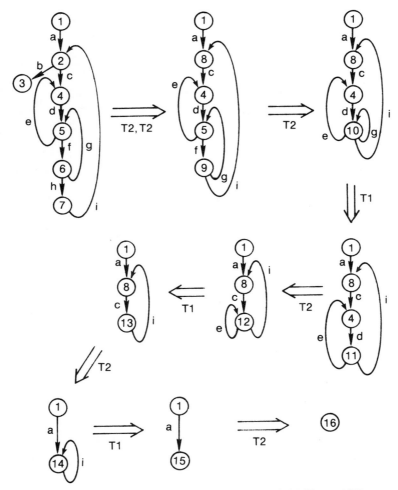

Figure 4.10 Parse of a reducible flow graph by T1 and T2.

LEMMA 4.10 *Let G be an RFG, and let π be a parse of G. If (x,y) is a backward arc, then there is a path from y to x in G.*

Proof Since (x,y) is a backward arc, at some intermediate stage of the reduction of G there is a node z and arc (z,z) such that (z,z) represents (x,y). Furthermore, node z represents a region R of G headed by y and containing x. (If y did not head R, then there would be a path from s to y containing an arc, namely (x,y), not in R. This path is easily seen to violate the definition of 'region'.) Since R itself is a flow graph, there is a path from y to x in R, and hence in G. □

LEMMA 4.11 *If G is an RFG and $R = (N_1, A_1, y)$ is any region of G with $x \in N_1$ and $(x,y) \in A_1$, then $(x,y) \in B(G)$.*

Proof Since R is a region with header y containing x, then y dominates x (by Lemma 4.4). If $y = x$, then $(x,y) \in B(G)$ by Lemma 4.8.

Now suppose that $y \neq x$. If (x,y) were forward, then there would exist a region $R' = (N_2, A_2, h)$ containing x (see Figure 4.11). Since R' consumes a region containing y, it follows that h dominates y by Lemma 4.7. Either $x = h$ or $x \neq h$. If $x = h$, then x dominates y. This is a contradiction, since distinct nodes cannot mutually properly dominate each other (by antisymmetry). Thus, $x \neq h$. But then there is a loop-free path from outside (y) that does not pass through the header of R'. □

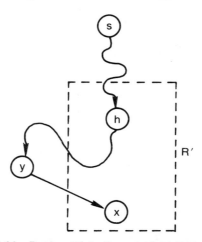

Figure 4.11 Region R' in the proof of Lemma 4.11.

DEFINITION If x and y are two not necessarily distinct nodes of a flow graph G, then we define the *distance from x to y*, denoted by $d(x,y)$, to be the minimum length of a path from x to y.

NOTATION If P and Q are paths of a flow graph, that is, ordered sets of nodes, then we let $P \triangledown Q$ stand for the set of nodes in both P and Q. In other words, we ignore the fact that P and Q are ordered and just take their unordered intersection.

THEOREM 4.6 *Let $G = (N, A, s)$ be an RFG and let π be a parse of G. Arc (x,y) is backward if and only if y dominates x.*

Proof *Only if.* Assume that (x,y) is backward. If $x = y$, then we are done. Suppose $x \neq y$. Then $x \neq s$ by Lemma 4.9. If $y = s$, then y dominates x by Lemma 4.3. If $y \neq s$, suppose that y does not dominate x. Let P be a

Regions, Parses, and Backward Arcs

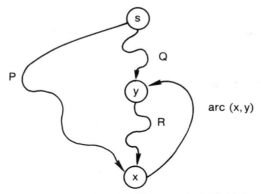

Figure 4.12 Paths in G in the proof of Theorem 4.6.

path from s to x not containing y. Let R be a path from y to x (by Lemma 4.10). Let Q be a path from s to y. Figure 4.12 shows this situation, where the paths shown are not necessarily disjoint.

Let i be a node from $P \nabla Q$ such that, if k is any other node from $P \nabla Q$ then $d(i,y) \leq d(k,y)$. Similarly, let j be a node from $P \nabla R$ such that, if k is any other node from $P \nabla R$, then $d(y,j) \leq d(y,k)$. (Note that s is in $P \nabla Q$ and that x is in $P \nabla R$, so that both intersections are nonempty. Furthermore, Q and R are arc disjoint and $Q \nabla R = \{y\}$.) But nodes i,j,y correspond to nodes a,b,c of (*), and thus G is irreducible, a contradiction. Thus, y dominates x.

If. Assume that y dominates x. If $x = y$, then (x,y) is backward by Lemma 4.8. Now suppose that y properly dominates x. There is a region $R = (N_1, A_1, y)$ in G where $x \in N_1$ and $(x,y) \in A_1$. In particular, $N_1 = \{i \in N | y$ dominates $i\}$ and $A_1 = A \cap (N_1 \times N_1)$ specifies one such R. But by Lemma 4.11, (x,y) is in $B(G)$ and hence backward in every parse π of G. □

Example 4.7 In Figure 4.13(i) we show an RFG G. It is easy to verify by using the dominance tree of G in Figure 4.13(ii) that for each backward arc (x,y) of G, y dominates x.

In the second part of Theorem 4.6 we have really shown that if $G = (N, A, s)$ is an RFG and if y dominates x in G, then $(x,y) \in B(G)$. Thus, we have as an immediate result that the backward arc set of an RFG is independent of the parse chosen and is $B(G)$.

COROLLARY 4.3 *If G is an RFG and if $B'(G)$ is the set of backward arcs produced by some particular parse of G, then $B(G) = B'(G)$. That is, the backward arcs of an RFG are unique.*

Characterizations of Reducible Flow Graphs

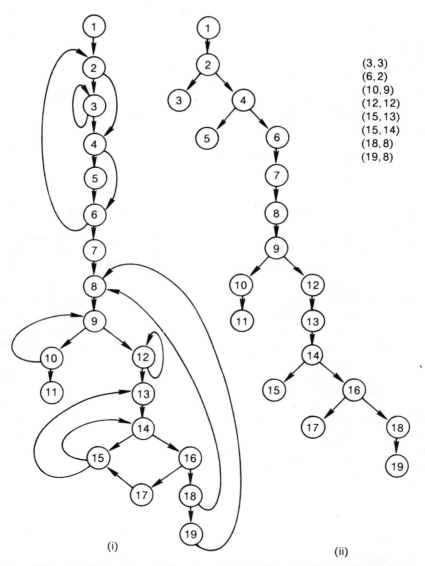

Figure 4.13 An Illustration of Theorem 4.6. (i) Flow graph G; (ii) dominance tree of G.

4.4 The DAG of a Reducible Flow Graph

In this section we show that all and only the RFGs can be decomposed uniquely into a DAG (directed acyclic graph) and backwards arcs. This DAG is exactly the DAG formed by a depth-first spanning tree of G including its forward and cross arcs but excluding its back arcs. Also, the dominance relation of an RFG and its DAG are the same.

DEFINITION A *DAG of a flow graph* $G = (N, A, s)$ is an acyclic flow graph $D = (N, A', s)$, such that $A' \subseteq A$ and for any arc e in $A - A'$, $(N, A' \cup \{e\}, s)$ is not a DAG. That is, D is a maximal acyclic subflowgraph.

Example 4.8 The solid arcs in the RFG of Figure 4.14 are its forward arcs and the dashed arcs are its backward arcs. The forward arcs form a DAG. Addition of any backward arc creates a cycle.

LEMMA 4.12 *If $G = (N, A, s)$ is an RFG, then $D = (N, A - B(G), s)$ is a DAG of G.*

Proof First we claim that D is acyclic. Let $\pi = X_1 X_2 \ldots X_m$ be a parse of G. The following inductive hypothesis is easy to prove by induction on i.

(IH): Let x be a node in the graph formed from G by steps $X_1 \ldots X_i$. Then there are no cycles among the forward arcs represented by X.

Taking IH for $i = m$ yields the claim.

Clearly D is a subflowgraph of G. Also, D is maximal, in that the addition of any back arc creates a cycle (by Lemma 4.10). □

THEOREM 4.7 *A flow graph $G = (N, A, s)$ is reducible if and only if its DAG is unique.*

Proof *Only if.* Assume that G is reducible but that its DAG is not unique. Then G has at least two DAGs, say $D' = (N, A', s)$ and $D'' = (N, A'', s)$. By Lemma 4.12, there is no loss of generality in assuming that $A' = A - B(G)$.

First, we note that $A' \neq A''$ since we assume that D' and D'' are distinct. Next, we observe that $A' \not\subseteq A''$ and $A'' \not\subseteq A'$ by the maximality property of a "DAG of a flow graph." Thus, there is an arc, say (x, y), in $A'' - A'$. Arc (x, y) is not a forward arc (removed by T2) because all forward arcs are in A'. Thus, (x, y) is in $B(G)$. If $x = y$, then D'' has a cycle and is not a DAG. So, $x \neq y$. Also, $x \neq s$ by Lemma 4.9. Since D'' is a flow graph, there is a path P from s to x in D''. If $y = s$, then P and arc (x, y) form a cycle. Thus, $y \neq s$. P does not contain y, for otherwise D'' is cyclic. But since (x, y) is in $B(G)$ and $x \neq y$, it follows that y dominates x by Theorem 4.6. However,

92 **Characterizations of Reducible Flow Graphs**

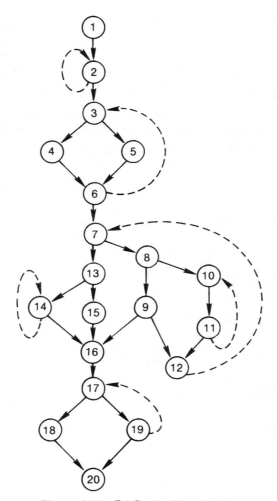

Figure 4.14 DAG of a flow graph.

the existence of a path P in G from the initial node to x not containing y is a contradiction.

If. Let G be a irreducible flow graph. We shall show that its DAG is not unique.

G contains at least one (*). Let V, W, X, Y, Z denote respectively the sets of arcs from the initial node to a, a to b, a to c, b to c, and c to b in one particular (*) of G. These five sets are disjoint, and V may be empty. (See Figure 4.15(i).)

We now construct two distinct DAGs, D' and D'', for G. Let G' be a DFST of G containing arcs $(V \cup W \cup Y \cup Z) - \{(d,b)\}$, where (d,b) is the

The DAG of a Reducible Flow Graph

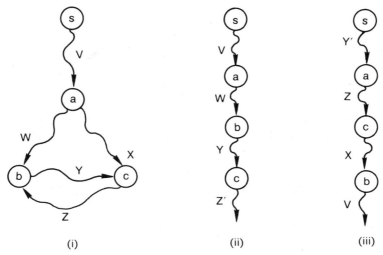

Figure 4.15 (i) The subgraph (*) in G. (ii) The beginning of D'. (iii) The beginning of D".

arc in Z that enters B. Figure 4.15(ii) shows how such a DFST can be started. $Z' = A - \{(d,b)\}$. Similarly, let G'' be a DFST of G containing arcs $(V \cup X \cup Z \cup Y) - \{(e,c)\}$, where (e,c) is the arc in Y that enters C. Figure 4.15(iii) shows how such a DFST can be started. $Y' = Y - \{(e,c)\}$. Let D' be G' plus its nonback arcs and let D'' be G'' plus its nonback arcs.

Clearly, both D' and D'' are DAGs of G. Also, neither one contains all the arcs in $Y \cup Z$, since this would yield a cycle. Each contains a different subset of the arcs of $Y \cup Z$. Thus, $D' \neq D''$. □

COROLLARY 4.4 *If $G = (N, A, s)$ is an RFG, then $D = (N, A - B(G), s)$ is its DAG.*

Proof It is a DAG of G (Lemma 4.12), and the DAG of G is unique (Theorem 4.7). □

COROLLARY 4.5 *The DAG of an RFG G is any DFST of G plus its forward and cross arcs. (Alternatively, the backward arcs of an RFG are exactly the back arcs of any DFST for G.) Thus, we can safely confuse backward arcs and back arcs in an RFG.*

Proof A DFST of G plus its forward and cross arcs is a maximal acyclic subflowgraph of G. Since G is an RFG, its DAG is unique (Theorem 4.7). □

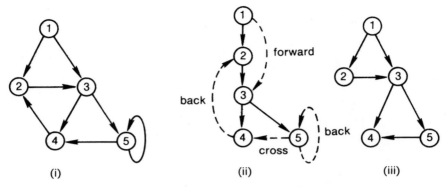

Figure 4.16 (I) RFG G. (II) DFST of G. (III) DAG of G.

Example 4.9 We show in Figure 4.16 that the DAG of an RFG G is the forward and cross arcs of a DFST of G.

COROLLARY 4.6 *A flow graph $G = (N, A, s)$ is reducible if and only if its arc set can be partitioned into two sets A_1 and A_2 such that $D = (N, A_1, s)$ is a DAG of G and for each (x, y) in A_2, y dominates x in G.*

Proof If G is reducible, let $A_2 = B(G)$.

If G is not reducible, it has a DAG other than D. Any DAG for G other than D would include an arc (x, y) such that y dominates x in G, an obvious impossibility. □

THEOREM 4.8 *A flow graph G is reducible if and only if its arc set can be partitioned into two sets A_1 and A_2 such that A_1 is a DAG D of G and each (x, y) in A_2 has y dominates x in D.*[6]

Proof By Corollary 4.6, it suffices to show that the dominance relations of D and G are the same.

If (x, y) is in $B(G)$ and y dominates x in G, then the removal of (x, y) will not change the dominance relation of the resulting graph.

Conversely, assume G is reducible and its arc set is partitioned into two sets A_1 and A_2 such that A_1 forms a DAG D and for each (x, y) in A_2, y dominates x in D. Let H be a list of all arcs (x, y) in A_2. Consider the sequence of flow graphs $D = G_0, G_1, \ldots, G_k = G$, where G_{i+1} is G_i plus the $(i+1)$st arc from H.

---

[6]Note the distinction between Theorem 4.8 and Corollary 4.6. Here the dominance relation is calculated in D rather than in G.

Loop-Connectedness and Interval Derived Sequence Length 95

INDUCTIVE HYPOTHESIS: If (x,y) is in H and u dominates v in D, then u dominates v in G_i.

BASIS: ($i=0$). Immediate.

INDUCTIVE STEP: ($i>0$). Assume the inductive hypothesis and consider G_i. Let (x,y) be the ith arc in H. If $x=y$, we are done. So, assume $x \neq y$, and suppose in contradiction that u dominates v in G_{i-1} yet u does not dominate v in G_i. Then $u \neq s$. Also, there exists a cycle-free path from s to v in G_i that does not contain u but must include arc (x,y). But then y does not dominate x in G_{i-1}, a contradiction. □

4.5 Loop-Connectedness and Interval Derived Sequence Length

In this section we establish a relationship between two parameters of an RFG. These parameters appear in the time complexity of some data flow analysis algorithms.

Initially, the loop-connectedness of a flow graph G WRT a DFST T of G was defined (in Chapter 3) as the largest number of back arcs (WRT T) found in any cycle-free path of G. But in an RFG, back arcs are backward arcs by Corollary 4.5. Thus, loop-connectedness of an RFG is independent of spanning tree concepts.

DEFINITION The *loop-connectedness of an RFG* G, which we shall denote by $d(G)$ or simply d, is the largest number of backward (or back) arcs found in any cycle-free path in G.

The other parameter of interest is the interval derived sequence length.

DEFINITION Let G be a flow graph, let $I(G)$ be the derived flow graph of G, and let G' be G minus all of its self-loops. (Recall that a self-loop is an arc from a node to itself.) We define the *length k of the derived sequence* of G to be 0 if G' is the trivial flow graph, otherwise that $k \neq 0$ such that

(a) $G_0 = G'$,
(b) $G_{i+1} = I(G_i)$, $i \geq 0$,
(c) G_k is the limit flow graph of G, and
(d) $G_k \neq G_{k-1}$.

The length of the derived sequence of an RFG corresponds intuitively to the maximum "nesting depth" of the loops of a reducible computer program. The "connectedness" of the loops of a computer program can be entirely different.

Example 4.10 Figure 4.17 shows the flow graphs of three nested **for**-loops and three nested **while**-loops with the corresponding k and d values. Figure 4.18 shows how the flow graph in Figure 4.17(ii) was obtained.

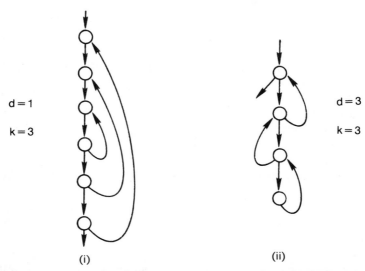

Figure 4.17 Loop-connectedness and interval derived sequence length. (i) Three for-loops. (ii) Three while-loops.

LEMMA 4.13 *Let G be an RFG and let G' be G at some intermediate stage of its reduction by T1 and T2. If there is a path from node u to node v in G', then there exist nodes w and x in G such that w and x are respectively represented by nodes u and v in G' and there is a path from w to x in G.*

Proof Let π be any parse of G that yields G' at some intermediate stage. The lemma is an easy induction on the number of steps of π taken to reach G'. □

LEMMA 4.14 *Let G be an RFG. Nodes entered by back arcs in G head intervals in G.*

Proof The lemma is obvious for self-loops. Therefore, let (m,h) be a back arc in G and suppose that $m \neq h$. Thus, h dominates m by Theorem 4.6. If h is the initial node, the lemma follows. Now consider the case where h is not the initial node.

Suppose h is in interval K but does not head K. Then by the method of constructing intervals (see Section 3.3.1), we must conclude that m is in K since (m,h) is an arc.

Since (m,h) is an arc, m must be added to K before h is. But then there is a path from the initial node to the header of K and then to m, which does not pass through h. This would contradict the assumption that h dominates m. □

Loop-Connectedness and Interval Derived Sequence Length

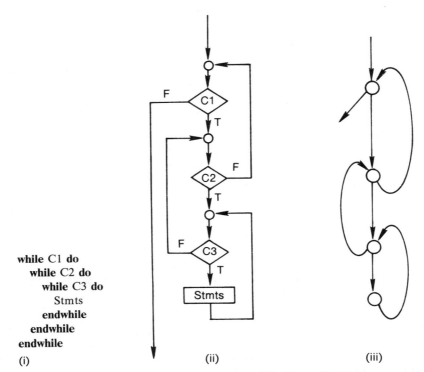

```
while C1 do
   while C2 do
      while C3 do
         Stmts
      endwhile
   endwhile
endwhile
```
(i)   (ii)   (iii)

Figure 4.18 The metamorphisis of three while-loops. (i) Pidgin SIMPL. (ii) Flow chart. (iii) Flow graph.

LEMMA 4.15 *If u dominates v in an RFG G, u heads an interval in G, J is the interval containing v, and $I(u) \neq J$, then $I(u)$ dominates J in $I(G)$.*

Proof Neither T1 nor T2 creates any new paths between nodes. That is, arcs of $I(G)$ are based precisely on arcs of G. Thus, if $I(u)$ did not dominate J, then u would not dominate v. □

Here is the key lemma for the theorem that follows.

LEMMA 4.16 *Let d be the loop-connectedness of an RFG G, let d' be the loop-connectedness of $I(G)$, and suppose that $G \neq I(G)$. Then $d' \geq d - 1$.*

Proof Assume all the hypotheses and let P be any cycle-free path in G from p_1 to p_m containing d back arcs. We write P as an ordered sequence of arcs, $P = [(p_1,p_2), (p_2,p_3), \ldots, (p_{m-1},p_m)]$, where the jth arc in P is (p_j, p_{j+1}). Let $[(x_1,r_1), (x_2,r_2), \ldots, (x_d,r_d)]$ be the subsequence of P consisting of all and only the back arcs of P. (See Figure 4.19.) Note that the r_i's are distinct, since P has no cycles.

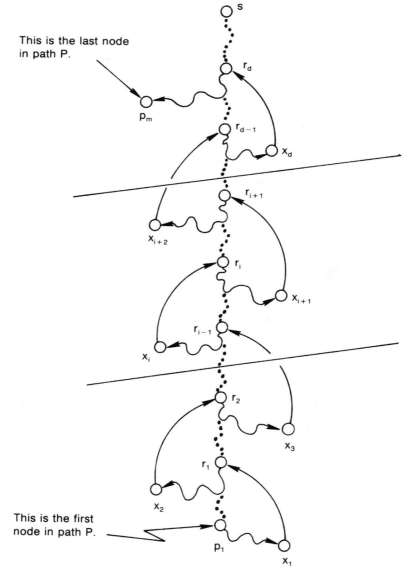

Figure 4.19 A cycle-free path P in an RFG from p_1 to p_m containing $d > 0$ back arcs.

Single-Entry Conditions

Let J_y denote the interval containing the node y. By Lemma 4.14, each r_i heads an interval J_{r_i}. Also, there is a path of non-back arcs from each r_i to x_{i+1}, so x_{i+1} cannot be in the interval $J_{r_{i+1}}$. Thus we have $J_{x_{i+1}} \neq J_{r_{i+1}}$, and the arc $(J_{x_{i+1}}, J_{r_{i+1}})$ is in $I(G)$. Furthermore, we know that r_{i+1} dominates x_{i+1} by Theorem 4.6 and so by Lemma 4.15 $J_{r_{i+1}}$ dominates $J_{x_{i+1}}$ in $I(G)$. Now, by Theorem 4.6 again, we may conclude that $(J_{x_{i+1}}, J_{r_{i+1}})$ is a back arc of $I(G)$. That is, each back arc of G, except possibly the first, is preserved in $I(G)$. If the first is also preserved, then $d' = d$. Otherwise, $d' = d - 1$. □

THEOREM 4.9 *If G is an RFG with loop-connectedness parameter d and derived sequence length k, then $k \geq d$.*

Proof Let $d = d_0, d_1, \ldots, d_k$ be the loop-connectedness parameters of the members of the derived sequence. By Lemma 4.16 we know that $d_{i-1} \leq d_i + 1$ for $1 \leq i \leq k$. Also, $d_k = 0$, since the last graph in the derived sequence is trivial. An elementary induction on i shows that $d_{k-i} \leq i$. Thus, $d_0 \leq k$, as was to be proved. □

4.6 Single-Entry Conditions

One general criterion proposed for well-formedness and easy readability of computer programs is that no "loop" have multiple entry points. In this section we show that reducible flow graphs have certain "single-entry" properties.

DEFINITION An *entry node of a cycle* C in a flow graph G is a node v in C entered by an arc (u, v) such that u is not in C. If C contains the initial node of G, then the initial node is also an entry node of C. We call a cycle with one entry node a *single-entry cycle*.

LEMMA 4.17

(a) *All flow graphs with only single-entry cycles are reducible.*
(b) *If a cycle is single-entry, then its entry node dominates all nodes in the cycle.*
(c) *If C is a cycle and v a node of C that dominates all nodes in C, then v is an entry node of C.*
(d) *If all strongly connected regions of a flow graph G are single-entry, then G is reducible.*

Proof (a) No such flow graph can contain (*). (b) By the definition of 'entry node of a cycle'. (c) By the definitions of 'flow graph' and 'entry node of a cycle'. (d) By Theorem 4.4. □

Recall that a simple cycle is a cycle $(x_1,\ldots,x_k)$, where $x_1,\ldots,x_{k-1}$ are distinct nodes.

THEOREM 4.10 *A flow graph G is reducible if and only if for each simple cycle C of G there is an entry node of C that dominates all other nodes in C.*

Proof *If.* If G is irreducible and we let C be the simple cycle in (*) from b to c and from c to b, then it is easy to see that no node of C dominates all other nodes of C. Note that C is a simple cycle since the paths involved are arc disjoint.

Only if. Let G be an RFG and let C be a simple cycle of G. There must be a back arc (x,y) in C, where $x \neq y$. Then y dominates x by Theorem 4.6. Thus, y is an entry node of C by Lemma 4.17. Suppose there were some node $z \neq y$ of C such that y did not dominate z. Then there would be a path from the initial node to z then along C to x which does not pass through y. Since y dominates x, it follows that y must dominate z. Thus, y is the desired entry node. □

DEFINITION Let $G=(N,A,s)$ be an RFG and let (x,y) be a backward arc of G. We define the *loop of backward arc (x,y)* to be the subflowgraph $L=(N',A',y)$ such that:

(a) N' contains x,y and all nodes z such that there is a path from z to x that does not pass through y.
(b) $A' = A \cap (N' \times N')$.

In an RFG with backward arc (x,y), we refer to the 'loop of (x,y)' by just the word 'loop'.

DEFINITION We define an *entry node of a loop* and a *single-entry loop* exactly as we did for a cycle, except that we replace each occurence of the word 'cycle' in that definition by the word 'loop'.

THEOREM 4.11 *In an RFG, the loop of backward arc (x,y) is single-entry.*

Proof The loop of (x,y) is a region with header y. □

There is another possible definition of 'entry node' and 'single-entry'.
DEFINITION Let $G=(N,A,s)$ be a flow graph, and let S be a subgraph of G. A node x of S is called an *entry node* of S iff there is a path from s to x containing no other nodes of S.

Example 4.11 Consider the subgraph of the graph in Figure 4.20 consisting of nodes 2, 3 and arcs $(2,3), (3,2)$. By the definition of entry node (of a

Subclasses of Reducible Flow Graphs

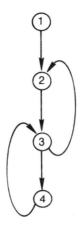

Figure 4.20 What is an entry node?

cycle), both nodes 2 and 3 are entry nodes. By the second definition of entry node, only node 2 is an entry node of the subgraph.

COROLLARY 4.7 *A flow graph G is reducible if and only if all strongly connected regions of G are single-entry (according to our second definition of 'entry node').*

Proof Both directions, in contrapositive form, follow immediately from the (*)-characterization (Theorems 4.4 and 4.5). □

4.7 Subclasses of Reducible Flow Graphs

The fact that certain classes of graphs are reducible is easily determined by using Theorem 4.8; namely, $G=(N,A,s)$ is an RFG if and only if A can be partitioned into two sets A_1 and A_2 such that $D=(N,A_1,s)$ is a DAG of G and for each $(x,y) \in A_2$, y dominates x in D.

One technique for showing the reducibility of a class of graphs is to define the class by an informal "graph grammar" and then observe that application of any "production" of this grammar preserves reducibility à la Theorem 4.8. We now consider an example of this.

Example 4.12 Figure 4.21 contains an informal graph grammar for a class of flow graphs (actually, flow charts) produced from one definition of 'structured programs'. Nodes are depicted by ovals, rectangles, diamonds, and circles. A rectangle containing an 'E' is an "expandable" node, whereas all other nodes are "non-expandable". Any expandable node can

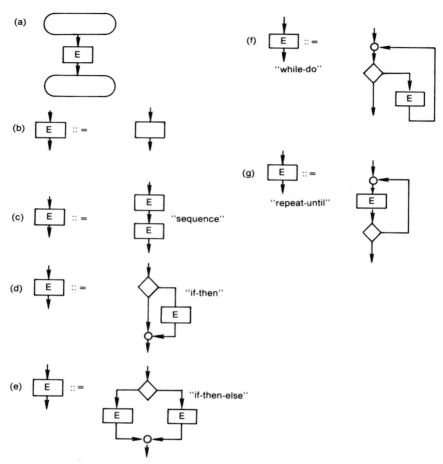

Figure 4.21 An informal graph grammar for a class of flow graphs.

be replaced by the right side of a production. To generate a flow graph by this grammar, start with Figure 4.21(a) and replace expandable nodes by right sides until there are no more expandable nodes. Such flow graphs are clearly reducible.

Here are some other subclasses of reducible graphs.

Example 4.13 The class of graphs generated by the grammar in Figure 4.21 and including **case** statements and "multi-level" **exit** statements is reducible.

Example 4.14 The flow graphs of those FORTRAN programs whose transfers to previous statements are all caused by normal termination of **DO**-loops are reducible.

Exercises

4.1 Can you think of other problems that employ finite Church-Rosser transformations?

4.2 Design an efficient algorithm to compute the loop of each backward arc of an RFG.

4.3 Design an $O(n)$, bottom-up, parsing algorithm for the grammar in Figure 4.21, where n is the number of nodes in the input graph.

4.4 Design an efficient algorithm to parse a proper program into its prime subprograms, where the terms "proper program" and "prime subprogram" are defined in Linger and Mills [1977]. (Hint: Define and use "forward dominance" and "back dominance".)

Bibliographic Notes

Transformations T1 and T2 and the equivalence of collapsibility and reducibility are from Hecht and Ullman [1972]. Various authors have considered similar transformations, but from the point of view of generating graphs rather than analyzing (i.e., reducing) them. For example, Cooper [1971] considers three generating rules, one of which is the inverse of T1 (i.e., addition of self-loops), and the other two together are equivalent to T2. In Engeler [1971a] and Engeler [1971b], "normal form flow charts" are considered that are built by two generating rules. One rule is the inverse of T1, and the other rule is equivalent to the inverse of T2 restricted so that the two nodes involved have disjoint sets of successors. Thus, normal form flow charts are characterized as trees with additional back arcs, a subclass of the reducible graphs.

Finite Church-Rosser transformations are discussed in Newman [1942]; Aho, Sethi, and Ullman [1972]; and Sethi [1974].

Transformations similar to T1 and T2 are discussed in Graham and Wegman [1976].

The (*)-characterization is from Hecht and Ullman [1972]. Schaefer [1973] and Allen and Cocke [1972] contain a result weaker than the (*)-characterization stating that every irreducible flow graph has a multi-entry cycle.

The concepts of "region", "representation", "parse", "DAG of an RFG", and Lemmas 4.5, 4.6, 4.12 come from Ullman [1973]. Theorems 4.6–4.10 come from Hecht and Ullman [1974, 1975]. Kasyanov [1973] contains two characterizations of reducibility: one in terms of single-entry strongly connected regions, and the other in terms of "node orders".

Experiments testing the incidence of reducibility are cited by Allen [1970] and Knuth [1971]. Various subclasses of reducible graphs are discussed in Bohm and Jacopini [1966]; Dijkstra [1968]; Cooper [1968]; Peterson, Kasami, and Tokura [1973]; Kosaraju [1974]; Kam [1973]; Kennedy [1975]; and Linger and Mills [1977], among many others. A discussion of graph grammars can be found in Pavlidis [1972] and Farrow, Kennedy, and Zucconi [1976].

Gannon and Hecht [1977] give an $O(n^3)$ algorithm for parsing a proper program into its prime subprograms.

Chapter 5

NODE ORDERS, NODE LISTINGS

One way to solve an intraprocedural data flow analysis problem is as follows:

(a) Find an equation relating information at the top (bottom) of each node in a flow graph with information at the top (bottom) of adjacent (predecessor or successor) nodes.
(b) Pick a strategy that specifies in what order nodes are to be "visited" such that, at completion of processing in step (d), there is no more information to be propagated.
(c) Select an appropriate initial estimate for the information posted at the top (bottom) of each node.
(d) Visit the nodes according to the strategy chosen in step (b). To visit a node, apply the equation found in step (a).

In this chapter we present two strategies that specify in what order nodes are to be visited in the above process. (We shall present the algorithms that use these ideas in Chapters 7 and 9.)

One strategy is to select a "reasonable node order" that includes each node exactly once, and visit the nodes in that order in round-robin fashion until all information being propagated "stabilizes."

Another strategy, called a "node listing," specifies a list of nodes where each node may occur more than once in the node listing, and only one pass through the nodes in the listing suffices for propagating information.

The round-robin method is very easy to program. However, each node is visited on each iteration and this is, in general, unnecessary. Also, a stabilization test is required after each iteration. The node listing method avoids these drawbacks, although the computation of efficient node listings is nonobvious in general.

One practical consequence of rPOSTORDER (one "reasonable node order") is a simple dominator algorithm for reducible flow graphs (given in Section 5.2).

5.1 Reasonable Node Orders

DEFINITION Let $G = (N, A, s)$ be a flow graph. A *node order* of G is a list of nodes that includes every node of G exactly once.

DEFINITION A node order R of a flow graph G is called *reasonable*[1] iff R topsorts the dominance relation of G. In addition, if G is an RFG, then R topsorts the DAG of G.

We shall now show that interval order and rPOSTORDER are reasonable. In Section 3.3.4 (on Interval Order) it was shown by an interval-theoretic proof that interval order topologically sorts the dominance relation. This result will pop out of the following as an immediate corollary.

5.1.1 Interval Order Is Reasonable

DEFINITION Let G be an RFG and let π be a parse of G by T1 and T2. A *reduction of an RFG G WRT a parse* π is any G_i, $0 \leq i \leq m$, where $G_0 = G$, G_{i+1} is the result of the $(i+1)$ st transformation in π applied to G_i, and G_m is the trivial flow graph.

DEFINITION The *string of a node* of a reduction of an RFG G WRT a parse π is defined recursively as follows:

(a) In G, the string of each node w is `w`.
(b) Whenever node x (with string α) consumes node y (with string β) in G_i to form G_{i+1}, then α`,`β is the string of the merged node z.
(c) The string of any node v, where v is not represented by x or y in (b), is the same in G_{i+1} as it was in G_i.

LEMMA 5.1 *In any reduction G' of an RFG G (WRT a parse π), the set of strings of nodes of G' is a partition of the nodes of G.*

Proof Let x be a node in G'. Node x represents a region R of G. Clearly, the string of x is the set of nodes in R. Disjointness follows from Lemma 4.5(c). By a straightforward induction on the number of steps in π, these strings collectively include all the nodes of G. □

DEFINITION Let x be a node in a reduction G' of an RFG G WRT parse π, and suppose x represents region R of G. A *parse order of a region* R is the string of x. A *parse order of an RFG G* is the string of the node representing the limit flow graph of G.

LEMMA 5.2 *If P is a parse order of an RFG G, then P is parse order of the DAG D of G.*

---

[1] The selection of the word 'reasonable' here, as opposed to some other word, was arbitrary. I think the word 'reasonable' is appropriate (i.e., 'reasonable' is reasonable), since I can think of no better adjective to use here.

Reasonable Node Orders

Proof Assume that P is a parse order of G. Clearly, each backward arc has no effect on the construction of P. Thus, all backward arcs can be ignored. Therefore, P is also a parse order of D. □

THEOREM 5.1 *Any parse order of an RFG G topsorts the DAG D of G.*

Proof
INDUCTIVE HYPOTHESIS: Let z be any node of any reduction of an RFG G WRT a parse π, let α be the string of node z, and let α have length k (i.e., α represents k nodes). String α topsorts the DAG of the region represented by z.
BASIS: ($k=1$). Obvious.
INDUCTIVE STEP: ($k>1$). Assume that the inductive hypothesis is true for all $i \leq k-1$. Suppose x (with string α) consumes y (with string β) to form node z with $\alpha\grave{},\grave{}\beta$ as its string. By the inductive hypothesis applied twice, α (respectively β) topsorts the DAG of the region that x (respectively y) represents. Thus, $\alpha\grave{},\grave{}\beta$ topsorts the region represented by z by the definition of T2. This ends the induction.
Consequently, applying the inductive hypothesis for $k = n$, the number of nodes in G, proves the theorem. □

COROLLARY 5.1 *An interval order topsorts the DAG of an RFG.*

Proof Since interval reduction can be simulated by T1–T2 reduction (Theorem 4.2), any interval order is a parse order. □

LEMMA 5.3 *Any linear order that topsorts the DAG D of an RFG G also topsorts the dominance relation of G.*

Proof If x properly dominates y, then x precedes y in D. Thus, the dominance relation is a subset of the "precedes" relation in D. Therefore, Observation 2.8 applies. □

COROLLARY 5.2 *An interval order topsorts the dominance relation of an RFG.*

Proof By Corollary 5.1 and Lemma 5.3. □

5.1.2 rPOSTORDER Is Reasonable

For convenience we employ the term POSTORDER as well as rPOSTORDER. POSTORDER is the order in which each node was last visited while growing a DFST of a flow graph. rPOSTORDER is the reverse of

POSTORDER. So, POSTORDER$[x] = n + 1 -$ rPOSTORDER$[x]$, for each node x.

THEOREM 5.2 *rPOSTORDER topsorts the DAG of an RFG. (That is, the partial order defined by the DAG of an RFG is a subset of the total order defined by rPOSTORDER.)*

Proof Let G be an RFG, let G' be the DAG of G, and let T be any DFST of G. It suffices to show that if there is a path in G' from the initial node to node y that includes node x, with $x \neq y$, then rPOSTORDER$[x] <$ rPOSTORDER$[y]$.

Suppose, in contradiction, that there are two distinct nodes x and y such that there is a path in G' from the initial node to y that includes x, and rPOSTORDER$[x] >$ rPOSTORDER$[y]$. Then POSTORDER$[x] <$ POSTORDER$[y]$. That is, y is last visited after x is last visited while growing T.

Either y is an ancestor of x, or y is "to the right" of x in T. If y is an ancestor of x, then G' contains a cycle. This is impossible. Consequently, y is to the right of x. The path from x to y must go through a common ancestor of x and y, so there would again be a cycle in G'. □

Now we have as an immediate corollary that rPOSTORDER topsorts the dominance relation of an RFG (via Lemma 5.3). But this result is subsumed in the next theorem where it is shown that rPOSTORDER topsorts the dominance relation of any flow graph—reducible or irreducible.

THEOREM 5.3 *rPOSTORDER topsorts the dominance relation of a flow graph. That is, if x dominates y, then rPOSTORDER$[x] \leq$ rPOSTORDER$[y]$.*

Proof Let G be a flow graph in which x dominates y, and let T be any DFST of G. The case where $x = y$ is easy, so suppose that $x \neq y$. Since any path from the initial node to y must include x, x is reached before y while growing T. Furthermore, x is an ancestor of y in T. If this were not so, then either y is an ancestor of x in T, or y is to the right of x in T, or x is to the right of y in T. Should one of these three be true, then there would be a path in G from s to y not containing x, contradicting x dominates y. Thus, y is last visited before x is last visited in Algorithm 3.3. That is, POSTORDER$[y] <$ POSTORDER$[x]$, or rPOSTORDER$[x] <$ rPOSTORDER$[y]$. □

5.2 A Dominator Algorithm for a Reducible Flow Graph

One consequence of a reasonable node order of an RFG is a fast bit vector step algorithm for computing dominators.

If x is a predecessor of y in an RFG, then either (x,y) is a back arc or it is not. If it is a back arc, then either y properly dominates x or $x=y$ (by Theorem 4.6 and Corollary 4.6), and thus x cannot properly dominate y. If (x,y) is not a back edge of an RFG, then rPOSTORDER$[x]<$ rPOSTORDER$[y]$.

This is exactly the property of rPOSTORDER that Algorithm 5.1 below uses.

ALGORITHM 5.1 *Computes sets $DOM(x)$, the dominators of x, for each node x.*

INPUT: $G=(N,A,s)$ is a reducible flow graph (with n nodes) represented by predecessor lists. The nodes of G are numbered from 1 to n by rPOSTORDER according to Algorithm 3.3. We refer to each node by its rPOSTORDER number.[2]

OUTPUT: Sets $DOM(y)=\{x|x \text{ dominates } y\}$, for each node y.

METHOD: Sets $DOM(1),\ldots,DOM(n)$ are global, and each is represented by a bit vector of length n.

See procedure FIND$DOMINATORS in Figure 5.1. □

---

procedure FIND$DOMINATORS (RFG $G=(N,A,s)$)
  integers j,k
  /*** Initialization ***/
  for $j := 1$ **to** n **by** 1 **do**
    $DOM(j) := \{j\}$
  endfor
  /*** One pass through nodes 2 to n ***/
  for $j := 2$ **to** n **by** 1 **do**
    $DOM(j) := \bigcap_{\substack{k \in \text{PRED}(j) \\ k<j}} [\{k\} \cup DOM(k)]$
  endfor
return

Figure 5.1 A dominator algorithm for an RFG.

[2]This assumption can be implemented in one of two ways. One way is to replace the statement "rPOSTORDER$[x] := i$" by "rPOSTORDER$[i] := x$" in Algorithm 3.3, and then use rPOSTORDER appropriately. Another way is to just use a "bucket sort" to renumber nodes.

THEOREM 5.4 *Algorithm 5.1 is correct. That is, after Algorithm 5.1 terminates, i is in DOM(j) if and only if i dominates j.*

Proof Let G be an RFG. We proceed by induction on j.

INDUCTIVE HYPOTHESIS: After processing node j, i is in DOM(j) if and only if i dominates j.

BASIS: ($j = 1$). Trivially true.

INDUCTIVE STEP: ($j > 1$). Assume the inductive hypothesis for all $k < j$, and consider the case for j.

If i dominates j, then surely i dominates every predecessor of j. Thus, i is in DOM(j).

Now, suppose i is in DOM(j), but i does not dominate j. Then there is a cycle-free path from the initial node to j that does not pass through i. Let k be the node on the path immediately before j. By Theorem 4.6, (k,j) cannot be a back arc, or else j would dominate k, and the path would have a cycle. Thus, (k,j) is not a back arc, and rPOSTORDER[k] < rPOSTORDER[j]. As $i \neq k$, and i does not dominate k, we have by the inductive hypothesis that i is not in $\{k\} \cup$ DOM(k), and hence, not in DOM(j). □

If the DOM sets are implemented by bit vectors as suggested, then Algorithm 5.1 requires $O(r)$ bit vector steps. This follows because in a flow graph with r arcs at most r bit vector intersections are computed in the second **for**-loop of Algorithm 5.1. Also, the node ordering (rPOSTORDER) assumed as input can be computed in $O(r)$ elementary steps.

5.3 Node Listings

When solving certain intraprocedural data flow analysis problems such as "reaching definitions," "live variables," and "available expressions,"[3] information at each node in a control flow graph that is propagated to all other nodes need only be propagated over cycle-free paths. (This will become evident in Chapter 7.) A "node listing" is an intermediate representation of a flow graph that facilitates this propagation.

There are two types of node listings: strong and weak. Strong node listings rely on the concept of a simple path, and weak node listings rely on the concept of a "basic" path. In Chapter 7 we shall see that problems such as "reaching definitions" and "live variables" may be solved with a weak node listing, however, "available expressions" requires a strong node listing.

---

[3]See Section 7.1.1 for a definition of the "available expressions" problem.

Node Listings

5.3.1 Strong Node Listings

DEFINITION A *strong node listing* for a flow graph $G = (N, A, s)$ is a sequence $(y_1, y_2, \ldots, y_m)$, $m \geq n$, of nodes from N (in which nodes may be repeated more than once) such that all simple paths of G are (not necessarily contiguous) subsequences thereof.

Example 5.1 $(1, 2, 3, 2, 1)$ is a strong node listing of the flow graph in Figure 5.2.

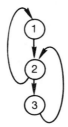

Figure 5.2 A flow graph.

OBSERVATION 5.1 *For any flow graph there exists a strong node listing of length at most $n(n-1) + 1$.*

Proof Consider the sequence of $n - 1$ repetitions of the nodes 1 to n in that order, followed by a 1. That is,

$$(1, 2, \ldots, n, \underbrace{1, 2, \ldots, n, \ldots, 1, 2, \ldots, n}_{n-1 \text{ repetitions of } (1,2,\ldots,n)}, 1).$$

Clearly, all permutations of $\{1, 2, \ldots, n\}$ are subsequences of this sequence. □

DEFINITION A strong node listing for a flow graph G is called *minimal* iff there is no shorter strong node listing for G.

Example 5.2 $(1, 2, 3, 2, 1)$ is a minimal strong node listing for the flow graph in Figure 5.2.

OBSERVATION 5.2 *If a flow graph $G = (N, A, s)$ is a DAG, then any topsort of G is a (minimal) strong node listing (of length n) for G.*

Three important results about strong node listings for RFG (whose proofs we shall omit) are:

(1) Every RFG has a strong node listing of length at most $n + 2.01 n \log n$.
(2) There exist RFGs for which no strong node listing is shorter than $\frac{1}{2} n \log n$.
(3) A strong node listing for an RFG with $r = O(n)$ arcs can be constructed in $O(n \log n)$ time.

5.3.2 Weak Node Listings

DEFINITION A *basic path* in a flow graph $G = (N, A, s)$ is a simple path $(x_1, x_2, \ldots, x_k)$, $k \geq 1$, such that there is no shorter simple path from x_1 to x_k which is a subsequence of $(x_1, \ldots, x_k)$.

Example 5.3 In Figure 5.3, the simple path $(1, 2, 3, 4, 5)$ is not basic because it contains the basic path $(1, 2, 4, 5)$.

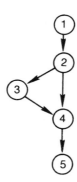

Figure 5.3 Simple and basic paths. Path $(1, 2, 3, 4, 5)$ is not basic; path $(1, 2, 4, 5)$ is basic.

DEFINITION A *weak node listing* for a flow graph $G = (N, A, s)$ is a sequence $(y_1, y_2, \ldots, y_m)$, $m \geq n$, of nodes from N (in which nodes may be repeated more than once) such that all basic paths of G are (not necessarily contiguous) subsequences thereof.

OBSERVATION 5.3 *Every strong node listing is also a weak node listing.*

A *minimal weak node listing* is defined analogously to a minimal strong node listing.

It has been shown that there exist RFGs for which no weak node listing is shorter than $\frac{3}{8} n \log n$. However, various classes of flow graphs corresponding to certain classes of programs have been shown to have weak node listings of length $O(n)$ by Kennedy [1975].

Bibliographic Notes

Node ordering is discussed in Earnest, Balke, and Anderson [1972]; Schaefer [1973]; Kasyanov [1973]; Hecht [1974]; and Hecht and Ullman [1975].

The dominator algorithm in Section 5.2 is from Hecht and Ullman [1975]. Other dominator algorithms can be found in Allen and Cocke [1972], Purdom and Moore [1972], and Aho and Ullman [1973]. Aho, Hopcroft, and Ullman [1973] give an $O(r \log r)$ step algorithm to compute direct dominators in an RFG. Tarjan [1973] presents an $O(r + n \log n)$ step algorithm to find direct dominators in an arbitrary flow graph.

Our treatment of node listings comes from Kennedy [1975]. Aho and Ullman [1975] present an $O(n \log n)$ algorithm to find a length $O(n \log n)$ strong node listing of an RFG, and point out the distinction between strong and weak node listings. Markowsky [1974] and Tarjan [1974] have independently shown that there exist RFGs for which no strong node listing is shorter than $\frac{1}{2} n \log n$. Aho and Ullman [1975] can show that Tarjan's construction also proves that $\frac{3}{8} n \log n$ is necessary even for weak node listings. Frederickson [1975] has found some improvements in Aho and Ullman's algorithm that shortens (by a constant factor) the length of the node listing found by their algorithm.

Chapter 6

NODE SPLITTING

"Node splitting" is the replication of certain nodes in a graph and generation of an "equivalent" graph. This process, or graph transformation, can be used to convert an irreducible flow graph (acronym: IRFG) into an RFG. But although this is the motivation for its discussion in this chapter, the concept of node splitting appears in several other areas.

Node splitting can be used in certain instances to transform "**go to** programs" into "**while** programs". It is also one of several possible transformations for converting flow graphs to certain standard forms to simplify the proving of theorems (such as correctness) about computer programs.

In this chapter we shall prove that for an IRFG, graph reduction (e.g., by T1 and T2, or by intervals) must be interleaved with node splitting in that all node splitting cannot, in general, precede reduction. This result is basically of theoretical interest in that it thwarts one potential strategy for minimizing the extent of splitting.

Known algorithms (e.g., Cocke [1971], and Allen and Cocke [1972]) for minimizing the amount of splitting first solve a "minimum covering problem" or a "minimum feedback arc problem". There are no known efficient algorithms for these problems.[1]

In practice, no or little node splitting is necessary. Nevertheless, the problem of irreducible flow graphs cannot be ignored, in general, by algorithms such as interval analysis.

6.1 One Simple Criterion

Since not all flow graphs are reducible, it is natural to ask what procedures exist for transforming an IRFG into an "equivalent" RFG.

Assume that the nodes of a flow graph have (not necessarily distinct) node labels. If $P = (x_1, \ldots, x_k)$ is a path in a flow graph, then we define labels(P) to be the string of labels of these nodes; that is, (label(x_1),...,label(x_k)). We say that two flow graphs G_1 and G_2 are *equivalent* iff, for each path P in G_1, there is a path Q in G_2 such that labels(P) = labels(Q), and conversely.

---

[1]They are "NP-Complete". See Aho, Hopcroft, and Ullman [1974], for example, for a definition of this term and definitions of the covering and feedback arc problems.

A Definition of Node Splitting

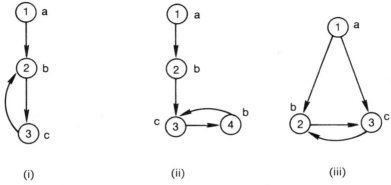

Figure 6.1 "Equivalence".

Example 6.1 Flow graphs (i) and (ii) of Figure 6.1 satisfy the definition of equivalence. Flow graphs (i) and (iii) of Figure 6.1 are not equivalent because there is no path in (i) that has label list (a,c).

6.2 A Definition of Node Splitting

DEFINITION Let $G=(N,A,s)$ be a flow graph. Let x ($x \neq s$) be a node with no self-loop, predecessors $w_1,\ldots,w_p$ ($p \geq 2$), and successors $y_1,\ldots,y_s$ ($s \geq 0$). We define $G \operatorname{split} x$ to be the flow graph resulting from the following procedure:

(a) Delete arcs $\{(w_i,x)|1 \leq i \leq p\} \cup \{(x,y_j)|1 \leq j \leq s\}$ and node x from G.
(b) Add p copies of x (called $x_1,\ldots,x_p$) to G. Add arcs $\{(w_i,x_i)|1 \leq i \leq p\} \cup \{(x_i,y_j)|1 \leq i \leq p, 1 \leq j \leq s\}$ to G.
(c) For each x_i, $1 \leq i \leq p$, label(x_i) := label(x). All other labels remain the same.

Node splitting as defined above is the process of splitting a node with one or more entering arcs (and no self-loop) into several identical copies, one for each entering arc. It is easy to verify that if x is split in flow graph G, then G is equivalent to ($G \operatorname{split} x$). Furthermore, since the "is equivalent to" relation is transitive, the result of splitting a finite number of nodes in either G or any of its equivalent graphs is equivalent to G.

There are several other definitions of node splitting. In one version, the predecessors of a node x (assuming x has no self-loop) are partitioned into two nonempty sets P and Q; all arcs from P to x are deleted; a new copy

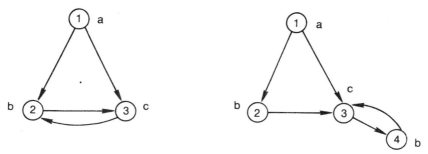

Figure 6.2 Converting an IRFG into an RFG.

of x called x' is added; and arcs from P to x' and from x' to each successor of x are added. This transformation is more general than our definition. Other studies have defined node splitting to include the replication of subgraphs of a flow graph.

Example 6.2 Figure 6.2 shows a simple example of the use of node splitting to convert an IRFG into an RFG. Note that T2 is always applicable after node splitting.

If we interleave the processes of interval or T1-T2 reduction with node splitting, then we can always transform a flow graph into a single node.

Let S denote the splitting of a node, and let $\hat{T}$ denote the completion of T1 and T2 (i.e., $\hat{\Rightarrow}$ as defined in Chapter 4).

THEOREM 6.1 Any flow graph can be transformed into the trivial flow graph by a transformation represented by the regular expression[2] $\hat{T}(S\hat{T})^*$. That is, first apply $\hat{T}$, then apply S followed by $\hat{T}$ zero or more times.

Proof Clearly, $\hat{T}$ always decreases the number of nodes by at least one (since T2 is applicable to each copy of the split node), and we only begin with a finite number of nodes. □

6.3 The power of Splitting by Itself

Suppose we have an IRFG. We can reduce G by a transformation of the form $\hat{T}(S\hat{T})^+$, where the '+' here indicates one or more (but a finite number of) applications of $S\hat{T}$. Can we reduce G by using an operator of the form S^+T? Unfortunately, the answer to this is, in general, no.

---

[2]For a definition of regular expressions and the sets they denote, see Aho and Ullman [1972], for example.

The Power of Splitting by Itself

We now define a certain property X such that a flow graph with property X is irreducible and, moreover, property X is preserved by node splitting. Hence, there exist irreducible flow graphs that cannot be transformed into reducible flow graphs by node splitting alone.

DEFINITION A flow graph has *property* X iff it contains at least five nodes a,b,c,d,e such that:

(1) Nodes a,b,c and the initial node form a (*) subgraph as defined in Chapter 4.
(2) Let $P(x,y)$ denote a path from x to y.
    $P(b,d)$, $P(d,b)$, $P(c,e)$, $P(e,c)$ exist.
    The paths of (*) involving nodes s,a,b,c and paths $P(d,b)$ and $P(e,c)$ are arc disjoint (exclusive of endpoints).
(3) Path $P(b,d)$ (respectively $P(c,e)$) is either arc disjoint from all other paths previously mentioned or is contained on path $P(b,c)$ (respectively $P(c,b)$).

DEFINITION Node d is called *inside* iff $P(b,d) \subseteq P(b,c)$ with d on $P(b,c)$. Otherwise, $P(b,d)$ and $P(b,c)$ are arc disjoint and d is called *outside*. Similarly, e is called *inside* iff $P(c,e) \subseteq P(c,b)$ with e on $P(c,b)$, and *outside* iff $P(c,e)$ and $P(c,b)$ are arc disjoint.

Example 6.3 Three examples of flow graphs enjoying property X are exhibited in Figure 6.3. These graphs, with arcs instead of paths, depict the "forbidden" subgraphs just defined.

THEOREM 6.2 *Splitting alone will not transform an IRFG with property X into an RFG. That is, $S^*\hat{T} \neq \hat{T}(ST)^*$.*

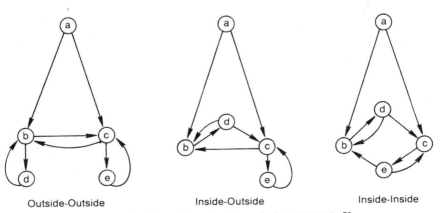

Outside-Outside   Inside-Outside   Inside-Inside

Figure 6.3 Some flow graphs enjoying property X.

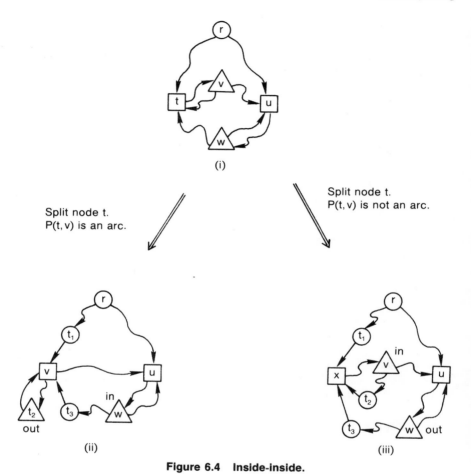

Figure 6.4 Inside-inside.

Proof We show by induction on the number k of splits that property X is preserved.

BASIS: ($k=0$). Trivial.

INDUCTIVE STEP: ($k>0$). Assume the inductive hypothesis is true for $k-1$ splits and consider the kth split. There are three cases to consider.

CASE 1: Inside-inside. The flow graph contains an inside-inside subgraph with property X as shown in Figure 6.4(i), where nodes r,t,u,v,w represent nodes a,b,c,d,e respectively.[3] Both t and u have at least two entering arcs, so each is a candidate to be split. (Clearly, if a node other

---

[3]In Figures 6.4, 6.5, and 6.6 the box shaped nodes play the role of nodes b,c, and the triangle shaped nodes play the role of nodes d,e.

The Power of Splitting by Itself

than t or u is split, property X is preserved.) By symmetry, splitting t is the same as splitting u. So we consider the case where t is split.

There are two subcases to consider. If $P(t,v)$ is the edge (t,v), then splitting t yields Figure 6.4(ii), and r,v,u,t_2,w represent a through e respectively. If $P(t,v)$ is not an arc, then let (t,x) be the first arc on this path. The result of splitting t is shown in Figure 6.4(iii), where r,x,u,v,w represent a through e respectively.

We continue in a similar manner for the next cases. Table 6.1 summarizes all three cases.

**Table 6.1
Summary of Cases in Proof of Theorem 6.2**

| | | | | | \multicolumn{5}{c}{NODE PLAYING ROLE OF} |
CASE	SPLIT	x	y	z	a	b	c	d	e
1(i)	—				r	t	u	v	w
1(ii)	t				r	v	u	t_2	w
1(iii)	t	✓			r	x	u	v	w
2(i)	—				r	t	u	v	w
2(ii)	t				r	v	u	t_3	w
2(iii)	t	✓			r	x	u	t_3	w
2(iv)	t		✓		r	v	u	t_3	w
2(v)	t	✓	✓		r	x	u	t_3	w
3(i)	—				r	t	u	v	w
3(ii)	t				r	v	u	t_3	w
3(iii)	t				r	x	u	v	w
3(iv)	u				r	t	w	v	u_3
3(v)	u		✓	✓	r	t	z	v	u_3
3(vi)	u		✓		r	t	w	v	u_3
3(vii)	u			✓	r	t	z	v	u_3

CASE 2: Outside-outside. See Figure 6.5.
CASE 3: Inside-outside. See Figure 6.6.

The three cases considered are summarized in Table 6.1. Cases 1(i), 1(ii) and 1(iii) in the table refer to their respective diagrams in Figure 6.4. In case 1(iii) corresponding to Figure 6.4(iii), we have chosen arc (t,x) as the first arc on path $P(t,v)$. The check mark in column x of row 1(iii) in Table 6.1 shows that node x is present; that is, path $P(t,v)$ is not an arc.

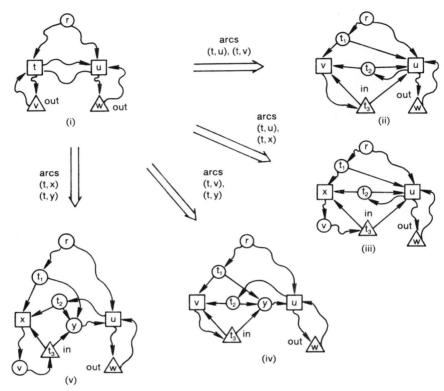

Figure 6.5 Outside-outside.

Similarly, Case 2 and Case 3 correspond to Figures 6.5 and 6.6.

The convention regarding nodes x, y and z is as follows. If $P(t,v)$ is not an arc, then (t,x) is the first arc on this path. Thus, x is present and this is recorded by a '$\sqrt{}$' in the proper place in Table 6.1. If $P(t,u)$ is not an arc, then (t,y) is the first arc on this path. Similarly, if $P(u,w)$ is not an arc, then (u,z) is the first arc on this path.

Finally, in all three cases, if a node other than those indicated in Table 6.1 is split, property X is clearly preserved. □

6.4 Why Node Splitting Is Not a Burning Issue

Certain intraprocedural data flow analysis algorithms (e.g., Cocke-Allen interval analysis, Ullman's algorithm, and the Graham-Wegman algorithm) require the input control flow graph to be reducible. If the input

6.4 Why Node Splitting Is Not a Burning Issue

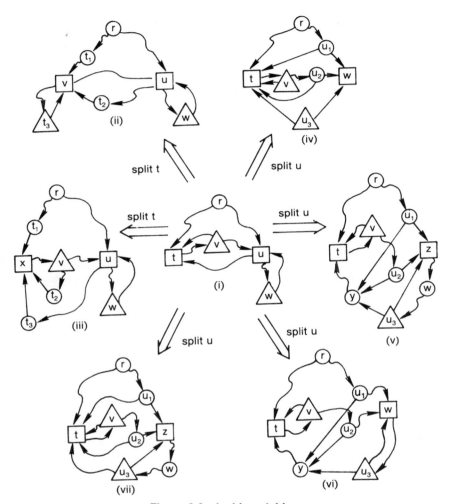

Figure 6.6 Inside-outside.

flow graph is irreducible, node splitting can be used to obtain an equivalent reducible flow graph.

The problems of how, what, and when to split are intriguing. However, the problem of irreducibility can be obviated in at least two ways. First, the iterative approach to intraprocedural data flow analysis (see Chapters 7 and 9), which is oblivious to the reducibility of the underlying flow graph, can be used. (In fact, we believe that it *should* be used.) Second, pervasive use of "disciplined" control flow structure that is clearly reducible (sometimes coerced by the programming language itself) is prevalent.

Exercises

6.1 Convert the Deutsch-Schorr-Waite marking algorithm, as presented in Section 2.3.5 of Knuth [1974], to a "structured program." Use your favorite defintion of a "structured program." It is unfair to use recursion here. Observe that this algorithm is essentially a nonrecursive implementation of depth-first search. Aim for clear readability.

6.2 Convert algorithms A, B, and C in Section 2.4 of Knuth [1974] to structured programs. You may use resursion if you desire here. I highly recommend recursion for C.

6.3 Convert the following algorithms from Section 6.4 of Knuth [1973] to structured programs:
  (a) Algorithm C—Collision resolution by Coalesced chaining.
  (b) Algorithm L—Collision resolution by Linear probing.
  (c) Algorithm D—Collision resolution by Double hashing.

6.4 Which of the following figures in Knuth [1974] are irreducible?
  (a) 38
  (b) 40
  (c) 41

6.5 Show that certain versions of the node splitting problem are NP-complete.

Bibliographic Notes

Node splitting for converting an IRFG into an RFG is discussed in Cocke and Miller [1969], Cocke and Schwartz [1970], Cocke [1970, 1971], Allen and Cocke [1972], Aho and Ullman [1973], and Schaefer [1973]. Theorem 6.2 is not published elsewhere.

Other uses of node splitting are discussed in Cooper [1968], Cooper [1971], Knuth and Floyd [1971], Ashcroft and Manna [1971], Kosaraju [1971], and Peterson, Kasami, and Tokura [1973].

Part III
Data Flow Analysis

Chapter 7

SOME SIMPLE ITERATIVE ALGORITHMS FOR DATA FLOW ANALYSIS

In this part of the book we shall present several intraprocedural data flow analysis algorithms that are based on flow graphs, and give some guidance for when to select a particular algorithm.

Several variations of an iterative algorithm (namely, Algorithm 2.2) are considered in this chapter: worklist versions, the round-robin version, and the node listing version. The worklist versions maintain a set containing "work-to-be-done" that is initialized, updated on-the-fly as the algorithm executes, and eventually exhausted. The worklist contains information to be propagated whose "influence" may not have been recorded yet. Nodes may be "visited" in an arbitrary order. The round-robin version propagates information by starting with an initial estimate of the desired information, then propagating information to nodes by repeatedly visiting the nodes in a round-robin fashion until information flow stabilizes (i.e., a fixed point is reached). The node listing version first preprocesses the flow graph to obtain a list of nodes (with repetitions, in general) then propagates information by visiting nodes in the order in which they occur on the list. The node listing has the property that visiting nodes in the indicated order suffices to propagate information.

The worklist and round-robin versions of the iterative algorithm have two important qualities that make them very desirable for practical use. First, these algorithms apply to all known data flow analysis problems. (See Chapter 9.) Second, these algorithms are very easy to program. No graph reductions are necessary (as in Cocke-Allen interval analysis, Ullman's algorithm, or the Graham-Wegman algorithm), and therefore the iterative algorithm is oblivious to the reducibility of the underlying flow graph. The disadvantage of the node listing version of the iterative algorithm is that the preprocessing necessary to compute a good node listing is usually nontrivial (cf. Aho and Ullman [1975], Kennedy [1975]).

There are undesirable qualities of the worklist and round-robin versions of the iterative algorithm. Most important is that fact that the worst case time complexity of these algorithms is not good. For example, with sparse reducible flow graphs on "bit vector problems" these algorithms require

$O(n^2)$ bit vector steps in the worst case, whereas Ullman's algorithm requires at most $O(n \log n)$ bit vector steps, and the Graham-Wegman algorithm requires at most $O(n)$ bit vector steps if the number of exits per loop is bounded. A second undesirable quality is that because the iterative algorithms do not analyze the underlying flow graphs, nothing is known about the loop structure of the graph, should this information be desired.

Nevertheless, we feel that ease of programming and generality make the iterative algorithm excellent for practical use when a flow graph is necessary. Consequently, we have included many versions of it in this chapter.

We present variations of the iterative algorithm applied to a class of very simple data flow analysis problems that are called "bit vector frameworks" in Chapter 9. Some reasons for considering just the bit vector frameworks here are: (1) it simplifies the exposition in that these problems are easier to understand than more general problems; and (2) such problems do occur often enough to justify a separate treatment.

7.1 Representative, Basic Data Flow Analysis Problems

There is an important subclass in intraprocedual data flow analysis problems each of which can be formulated as a collection of set equations, reminiscent of equations for conservation of flow, and the sets involved can be implemented by bit vectors. In general, a bit vector is associated with each flow graph node, each bit position corresponds to a program variable (or expression), and a bit indicates nonexistence or possible existence of an attribute of that variable (or expression) at that node.

We shall examine four representative problems of this subclass, two of which have already been introduced in Section 1.5.1. These problems are called

(1) "available expressions"
(2) "reaching definitions"
(3) "live variables", and
(4) "very busy expressions".

These problems are very similar in that almost any algorithm to solve one of these problems can, with slight modification, be used to solve the other problems. For this reason we use the "available expressions" and "live variables" problems to illustrate the algorithms in this chapter.

Since we restrict our attention here to intraprocedural rather than interprocedural problems, we assume that for the procedure under

Representative, Basic Data Flow Analysis Problems

scrutiny: (a) there is a control flow graph; (b) all relevant local data flow information is available; (c) all variable aliasing is known, can be handled, and thus for our purposes can be ignored; and (d) the procedure is isolated in that we may ignore where and how it is called.

As in Section 1.5.1, each problem below is described in terms of a flow graph. We associate certain information with the top of each node (pictorially following the "join" of the entry arcs), and other information with the bottom (pictorially preceding the "fork" of the exit arcs). The suffixes -TOP and -BOT distinguish these. The prefixes AE-, RD-, LV-, and VBE- are acronyms for the four problems.

7.1.1 Available Expressions

An expression such as $X + Y$ is *available* at a point p in a flow graph $G = (N, A, s)$ iff every sequence of branches that the program may take to p causes $X + Y$ to have been computed after the last computation of X or Y. By determining the set of available expressions at the top of each node in G, we know which expressions have already been computed prior to each node. Thus, we may be able to eliminate the redundant computation of some expressions within each node.

Example 7.1 Suppose the instructions in the flow graph in Figure 7.1(i) are translated into the quadruples in Figure 7.1(ii). Clearly, the expression $A + B$ is available at the top of nodes 2, 3, 4, and expression $C * D$ is available at the top of node 4. By appropriately renaming temporaries, Figure 7.1(ii) can be replaced by Figure 7.1(iii). This is the type of code improvement possible from knowing available expressions. Needless to say, there are other code improvements besides this one. For example, Figure 7.1(iv) shows how $C * D$ can be "hoisted" to node 1 from nodes 2 and 3.

Let E be the set of *expressions* defined in a flow graph. We assume that E is a nonempty finite set.

We assume also that there are two easily computed (local) functions from N to the power set of E. The first we call the NOTKILL function. NOTKILL(x) is interpreted as the set of expressions that are *not killed* in node x. Informally, expression $X + Y$ is *killed* iff the value of either X or Y may be modified within node x. The second we call the GEN function. If an expression such as $X + Y$ is in GEN(x), then we imagine that $X + Y$ is *generated* within node x; that is, $X + Y$ is evaluated within node x and neither X nor Y is subsequently modified within node x. Note that an

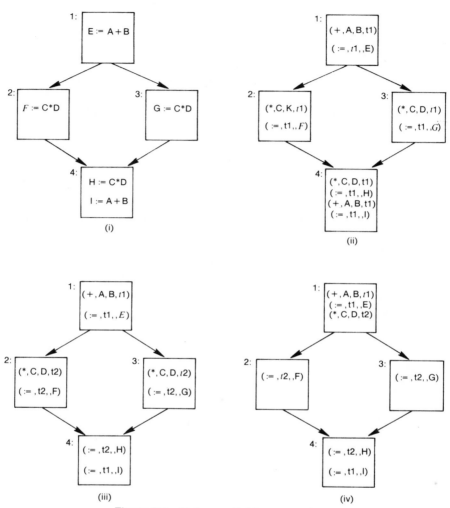

Figure 7.1 Using available expressions.

expression can be killed then generated in a node. We make the assumption that if an expression is killed somewhere, then it is generated somewhere.

Let AETOP(x) and AEBOT(x), for each node, x, be respectively the set of expressions available at the top of and bottom of node x.

The fundamental relationships that enable us to compute AETOP(x) for each node x are given in Figure 7.2. AEBOT can be similarly computed.

Representative, Basic Data Flow Analysis Problems

Equations for AETOP and AEBOT:
  AE1. $\text{AETOP}(s) = \emptyset$.
  AE2. For each node x, $x \neq s$, $\text{AETOP}(x) = \bigcap_{y \in \text{PRED}(x)} \text{AEBOT}(y)$
  AE3. For each node x,
    $\text{AEBOT}(x) = [\text{AETOP}(x) \cap \text{NOTKILL}(x)] \cup \text{GEN}(x)$.
  AE4. Since AE1–AE3 do not necessarily have a unique solution for AETOP, we desire the largest solution.
Equations for AETOP alone:
  AE1. $\text{AETOP}(s) = \emptyset$.
  AE5. For each node x, $x \neq s$,
    $\text{AETOP}(x) = \bigcap_{y \in \text{PRED}(x)} [(\text{AETOP}(y) \cap \text{NOTKILL}(y)) \cup \text{GEN}(y)]$.
  AE4. The largest solution of AETOP satisfying AE5 is desired.

Figure 7.2 Equations for "available expressions".

To see that the equations AE1–AE3 do not necessarily have a unique solution, consider the following example.

Example 7.2 In the flow graph in Figure 7.3, let $\text{AEBOT}(1) = \{X + Y\}$, $\text{NOTKILL}(2) = E$ (the set of expressions defined in the flow graph), and $\text{GEN}(2) = \emptyset$. What is $\text{AETOP}(2)$? $\text{AETOP}(2) = \text{AEBOT}(1) \cap [(\text{AETOP}(2) \cap \text{NOTKILL}(2)) \cup \text{GEN}(2)] = \{X + Y\} \cap [(\text{AETOP}(2) \cap E) \cup \emptyset] = \{X + Y\} \cap \text{AETOP}(2)$. Both $\text{AETOP}(2) = \emptyset$ and $\text{AETOP}(2) = \{X + Y\}$ are solutions to $\text{AETOP}(2) = \{X + Y\} \cap \text{AETOP}(2)$.

In Example 7.2, $\text{AETOP}(2) = \{X + Y\}$ is the correct answer. The reason for this hinges on the definition of the word 'available'. Since $\text{AETOP}(2)$ is supposed to be the set of definitions available at the top of node 2, and every sequence of branches that the program may take to the top of node 2 causes $X + Y$ to have been computed after the last computation of X and Y, then $\text{AETOP}(2) = \{X + Y\}$.

Figure 7.3 What is AETOP(2)?

7.1.2 Reaching Definitions

The "reaching definitions" problem has already been described in Section 1.5.1. Recall that we want to determine RDTOP(x), the set of definitions that reach the top of each node x. Figure 7.4 contains the basic equations for this problem, and equations for RDTOP alone.

Equations for RDTOP and RDBOT:
 RD1. For each node x, RDTOP(x) = $\bigcup_{y \in \text{PRED}(x)}$ RDBOT(y).
   (If PRED(s) = $\emptyset$, then RDTOP(s) = $\emptyset$.)
 RD2. For each node x,
   RDBOT(x) = [RDTOP(x) $\cap$ PRESERVED(x)] $\cup$ XDEFS(x).
 RD3. Since RD1 and RD2 do not necessarily have a unique solution for RDTOP, we desire the smallest solution.
Equations for RDTOP alone:
 RD4. For each node x,
   RDTOP(x) = $\bigcup_{y \in \text{PRED}(x)}$ [(RDTOP(y) $\cap$ PRESERVED(y)) $\cup$ XDEFS(y)].
   (If PRED(x) = $\emptyset$, then RDTOP(s) = $\emptyset$.)
 RD3. The smallest solution of RDTOP satisfying RD4 is desired.

Figure 7.4 Equations for "reaching definitions".

As with the "available expressions" problem, the solution to equations RD1–RD3 is not necessarily unique. However, this time the smallest (rather than largest) solution is required.

7.1.3 Live Variables

Figure 7.5 gives the basic equations for the "live variables" problem, which was also introduced in Section 1.5.1. Recall that LVBOT(x) is the set of (values of) variables which may be used after control passes the bottom of node x.

7.1.4 Very Busy Expressions

Let E be the set of expressions defined in a flow graph. We assume that E is a nonempty set. Let NOTKILL be as defined in the "available expressions" problem. In addition, we define a function XEUSES from N to the power set of E. XEUSES(x) is the set of expressions that have exposed uses in node x; that is, those expressions with a definition-clear path from the entry of node x to a use within node x. (XEUSES is analogous to XUSES of the "live variables" problem.)

Equations for LVTOP and LVBOT:
  LV1. For each node x, $\text{LVBOT}(x) = \bigcup_{y \in \text{SUC}(x)} \text{LVTOP}(y)$.
    (If $\text{SUC}(w) = \varnothing$ for any node w, then $\text{LVBOT}(w) = \varnothing$.)
  LV2. For each node x,
    $\text{LVTOP}(x) = [\text{LVBOT}(x) \cap \text{NOTDEFINED}(x)] \cup \text{XUSES}(x)$.
  LV3. The smallest solution for LVBOT satisfying LV1 and LV2 is desired.
Equations for LVBOT alone:
  LV4. For each node x,
    $\text{LVBOT}(x) = \bigcup_{y \in \text{SUC}(x)} [(\text{LVBOT}(y) \cap \text{NOTDEFINED}(y)) \cup \text{XUSES}(y)]$
    (If $\text{SUC}(w) = \varnothing$ for any node w, then $\text{LVBOT}(w) = \varnothing$.)
  LV3. The smallest solution of LVBOT satisfying LV4 is desired.

Figure 7.5   Equations for "live variables".

An expression is *very busy* at a point p in a flow graph iff it is always used before it is killed. Let $\text{VBETOP}(x)$ and $\text{VBEBOT}(x)$, for each node x, be respectively the set of expressions that are very busy at the top and bottom of node x.

The basic equations that enable use to compute $\text{VBEBOT}(x)$ for each node x are given in Figure 7.6.

Equations for VBETOP and VBEBOT:
  VBE1. For each node x, $\text{VBEBOT}(x) = \bigcap_{y \in \text{SUC}(x)} \text{VBETOP}(y)$.
    (If $\text{SUC}(w) = \varnothing$ for any node w, then $\text{VBEBOT}(w) = \varnothing$.)
  VBE2. For each node x,
    $\text{VBETOP}(x) = [\text{VBEBOT}(x) \cap \text{NOTKILL}(x)] \cup \text{XEUSES}(x)$.
  VBE3. The largest solution for VBEBOT satisfying VBE1 and VBE2 is desired.
Equations for VBEBOT alone:
  VBE4. For each node x,
    $\text{VBEBOT}(x) = \bigcap_{y \in \text{SUC}(x)} [(\text{VBEBOT}(y) \cap \text{NOTKILLED}(y)) \cup \text{XEUSES}(y)]$.
    (If $\text{SUC}(w) = \varnothing$ for any node w, then $\text{VBEBOT}(w) = \varnothing$.)
  VBE3. The largest solution for VBEBOT satisfying VBE4 is desired.

Figure 7.6   Equations for "very busy expressions".

7.1.5 A Taxonomy

Figure 7.7 summarizes the four types of basic data flow analysis problems. AE and VBE use the set intersection operation and require a largest solution, whereas, RD and LV use the set union operation and require a

	Set Intersection, "and", $\forall$-Problems	Set Union, "or", $\exists$-Problems
Top-Down Problems (operation over predecessors)	Available Expressions AE	Reaching Definitions RD
Bottom-Up Problems (operation over successors)	Very Busy Expressions VBE	Live Variables LV

Figure 7.7 The four types of bit vector frameworks.

smallest solution. For AE and RD the operation is over predecessors, whereas for VBE and LV it is over successors. We call AE and RD *top-down* (or forward) *problems* because information is propagated in the same direction as control flow to solve these problems. Conversely, we call VBE and LV *bottom-up* (or backward) *problems* because information must be propagated in the opposite direction of control flow to solve these problems.

Here are two observations about the previous four problems. Two definitons are needed for the first observation.

DEFINITION A *single-exit flow graph* is a quadruple $G=(N,A,s,t)$, where (N,A,s) is a flow graph, $t \in N$ is called *the exit node* of G, and there is a path from every node to t.

DEFINITION If $G=(N,A,s,t)$ is a single-exit flow graph, then $G^{-1}=(N,\{(y,x)|(x,y) \in A\},t,s)$ *is called* the reverse of G. (Note that G^{-1} is also a single-exit flow graph.)

OBSERVATION 7.1 *A top-down (respectively bottom-up) problem defined on a single-exit flow graph G can be transformed into a bottom-up (respectively top-down) problem on G^{-1} such that a solution to the modified problem gives an almost immediate solution to the original problem.*

Proof idea For example, convert a top-down problem on G that finds -TOPs to a bottom-up problem on G^{-1} that finds -BOTs. There is a simple relationship between -TOPs and -BOTs. □

OBSERVATION 7.2 *A set intersection (respectively set union) problem can be transformed into a set union (respectively set intersection) problem such that a solution to the modified problem gives a solution to the original problem.*

Proof idea Take set complements of the problem equations and use DeMorgan's Laws. □

7.2 Iterative Algorithm: Worklist Versions

We illustrate three worklist versions of the iterative algorithm by solving the "available expressions" problem. We dub these versions

(1) the segregated version,
(2) the integrated version, and
(3) Kildall's version.

In the segregated version we process one expression at a time. In the integrated version and Kildall's version we intermix the processing of expressions, but in slightly different ways.

First we discuss the segregated and integrated versions. Since the difference between them is trivial, we shall base our ensuing analysis on the segregated version.

Let $G = (N, A, s)$ be a flow graph in which nodes are numbered from 1 to n with $s = 1$, and let there be m expressions numbered from 1 to m. Let AETOPTABLE be an n-by-m array, where AETOPTABLE$[i,j]$ will ultimately be 1 if and only if expression j is in AETOP(i). During execution we may have AETOPTABLE$[i,j] = 1$ even though AETOP(i) does not contain expression j, but if AETOPTABLE$[i,j] = 0$, then expression j is certainly not in AETOP(i).

In the segregated version we process one expression at a time. To process an expression j, we find the set of nodes i such that expression j is not in AETOP(i). To do this we first initialize a worklist to $\{1\} \cup \{i|$ node i has a predecessor k such that expression j is killed in node k but not generated in node $k\}$. We then "propagate" 0's forward in the graph by setting AETOPTABLE$[k,j]$ to 0 iff it was previously 1, there is some predecessor i of k such that AETOPTABLE$[i,j] = 0$, and expression j is not generated by node i. The worklist contains nodes i for which there might possibly exist a k with i, j, k related as above. The AETOP information for each expression could be output on-the-fly rather than storing it as a table row-by-row.

In the integrated version we intermix the processing of expressions. In this version the worklist contains pairs (i,j) for which there might possibly exist a k with i, j, k again related as above.

ALGORITHM 7.1 *Segregated version of worklist approach.*

INPUT: (1) Flow graph $G = (N, A, s)$, $N = \{1, 2, \ldots, n\}$, $s = 1$. (2) Sets KILL(x) and GEN(x), for each node x, are implemented as bit arrays. There are m expressions numbered from 1 to m.

OUTPUT: AETOPTABLE, an n-by-m array, where AETOPTABLE$[i,j]$ = 1 iff expression j is in AETOP(i), and 0 otherwise.

METHOD: See procedure WORKLIST$VERSION1 in Figure 7.8. □

procedure WORKLIST$VERSION1
  queue Q, initially empty /* worklist */
  for $j := 1$ **to** m **by** 1 **do** /* for each expression do*/
    /* Step 1: Initialize worklist. */
    for $i := 1$ **to** n **by** 1 **do**
      if $i = 1$ $\vee$ node i has a predecessor k for
      which expression j is $KILL(k) - GEN(k)$ **then**
        AETOPTABLE$[i,j] := 0$
        Add i to Q.
      else
        AETOPTABLE$[i,j] := 1$
      endif
    endfor
    /* Step 2: Process worklist. */
    while Q is not empty **do**
      Let i be the first entry on Q.
      Remove i from Q.
      if expression j is not in GEN(i) **then**
        for each successor k of i **do**
          if AETOPTABLE$[k,j] = 1$ **then**
            AETOPTABLE$[k,j] := 0$
            Add k to Q.
          endif
        endfor
      endif
    endwhile
  endfor
  Output AETOPTABLE.
return

Figure 7.8 Segregated version of worklist approach.

ALGORITHM 7.2 *Integrated version of worklist approach.*
  INPUT: Same as for Algorithm 7.1.
  OUTPUT: Same as for Algorithm 7.1.
  METHOD: See procedure WORKLIST$VERSION2 in Figure 7.9. □

THEOREM 7.1 *Algorithms 7.1 and 7.2 terminate and are correct.*

Proof We consider only Algorithm 7.1. A similar argument applies for Algorithm 7.2.
Termination. For each expression j, no node i is placed on Q more than once and each iteration of the **while**-loop in step 2 of the algorithm removes an entry from Q.

Iterative Algorithm: Worklist Versions

procedure WORKLIST$VERSION2
  queue Q, initially empty /* worklist */
  /* Step 1: Initialize worklist. */
  for $i := 1$ **to** n **by** 1 **do**   /* for each node */
    for $j := 1$ **to** m **by** 1 **do**   /* for each expression */
      if $i = 1$ $\vee$ node i has a predecessor k for which
      expression j is in $\text{KILL}(k) - \text{GEN}(k)$ **then**
        $\text{AETOPTABLE}[i,j] := 0$
        Add (i,j) to Q.
      else
        $\text{AETOPTOBLE}[i,j] := 1$
      endif
    endfor
  endfor
  /* Step 2: Process worklist. */
  while Q is not empty **do**
    Let (i,j) be the first entry on Q.
    Remove (i, j) from Q.
    if expression j is not in $\text{GEN}(i)$ **then**
      for each successor k of i **do**
        if $\text{AETOPTABLE}[k,j] = 1$ **then**
          $\text{AETOPTABLE}[k,j] := 0$
          Add (k,j) to Q.
        endif
      endfor
    endif
  endwhile
  Output AETOPTABLE.
return

Figure 7.9   Integrated version of worklist approach.

Correctness. We must show that at termination, $\text{AETOPTABLE}[i,j] = 1$ iff expression j is in $\text{AETOP}(i)$.

Note that expression j is not in $\text{AETOP}(i)$ iff either

(a) there is a path of $k \geq 1$ nodes from s to i such that expression j is not generated in any node preceding i on this path, or

(b) there is a path of $k \geq 2$ nodes to i such that expression j is killed in the first node of this path and not subsequently generated.

If $\text{AETOPTABLE}[i,j]$ is set to 0 in step 1, clearly one of (a) or (b) holds. A straightforward induction on the number of iterations of the **while**-loop in step 2 shows that if $\text{AETOPTABLE}[i,j]$ is set to 0 in step 2, then (a) or (b) must again exist.

Conversely, if (a) or (b) holds, then a straightforward induction on k shows that AETOPTABLE$[i,j]$ is eventually set to 0. □

THEOREM 7.2 *Algorithms 7.1 and 7.2 each require at most $O(mr)$ elementary steps, where m is the number of expressions and r is the number of flow graph arcs.*

Proof For Algorithm 7.1, step 1 requires $O(n)$ steps. Step 2 requires at most $O(r)$ steps because each arc is considered (in a constant number of steps) at most once. Since there are m iterations of these two steps, and $r \geq n - 1$, we have $O(mr)$.

For Algorithm 7.2, steps 1 and 2 each require at most $O(mr)$ steps. This is because the aggregate number of predecessors (and successors) is $O(r)$. Thus, the total is $O(mr)$ steps. □

OBSERVATION 7.3 *To solve the "live variables" problem by the worklist approach, a table LBOTTABLE would be constructed by propagating 1's (instead of 0's).*

Here is Kildall's version of the worklist approach.

ALGORITHM 7.3 *Kildall's version of worklist approach.*
INPUT: Same as for Algorithm 7.1 except that NOTKILL(x), the complement of KILL(x), is used, and all NOTKILL and GEN sets are represented by bit vectors. There is a global array $A[1:n]$ of length m bit vectors.
OUTPUT: At the end of execution, $A[i]$ is the bit vector representation of AETOP(i), $1 \leq i \leq n$.
METHOD: See procedure WORKLIST$VERSION3 in Figure 7.10. The order in which pairs are selected from W in Step 2 is left unspecified. For example, W can be a queue or a stack. The AND (bitwise product) and OR (bitwise sum) functions are denoted by **bvand** and **bvor** respectively.
□

7.3 Iterative Algorithm: Round-Robin Version

In this section we present a round-robin version of the iterative algorithm to solve AE and LV. This version uses bit vectors and the bit vector operations **bvand** and **bvor**. Instead of maintaining a worklist, we repeatedly visit each node in round-robin fashion and propagate 0's forward for AE (1's backward for LV). The algorithm terminates when one iteration fails to change any bit of any bit vector.

ALGORITHM 7.4 *Round-robin version of iterative algorithm for AE.*
INPUT:

Iterative Algorithm: Round-Robin Version

```
procedure WORKLIST$VERSION3
  set W /* worklist */
  bit vector B of length m
  /* Step 1a: Initialize A. */
  A[1] := 0^m /* b^m denotes a bit vector of m b's */
  for i := 2 to n by 1 do A[i] := 1^m endfor
  /* Step 1b: Initialize W. */
  for each j ∈ SUC(1) do W := W ∪ {(j, GEN(1))} endfor
  for i := 1 to n by 1 do
    B := GEN(i) bvor NOTKILL(i)
    for each j ∈ SUC(i) do W := W ∪ {(j, B)} endfor
  endfor
  /* Step 2: Process W. */
  while W ≠ ∅ do
    Select and delete any pair (i, B) from W.
    if B has a 0 in some bit position in which A[i] has a 1 then
      A[i] := A[i] bvand B
      B := (A[i] bvand NOTKILL(i)) bvor GEN(i)
      W := W ∪ {(j, B) | j ∈ SUC(i)}
    endif
  endwhile
return
```

Figure 7.10 Kildall's version of worklist approach.

(1) Flow graph $G = (N, A, s)$, $N = \{1, 2, \ldots, n\}$, $s = 1$. The nodes are numbered from 1 to n by rPOSTORDER. We refer to each node by its rPOSTORDER number.

(2) Sets NOTKILL(i) and GEN(i), $1 \leq i \leq n$, represented by bit vectors of length m.

OUTPUT: Sets AETOP(i), $1 \leq i \leq n$, represented by bit vectors.

METHOD: See procedure AE$ROUND$ROBIN$VERSION in Figure 7.11. □

ALGORITHM 7.5 *Round-robin version of iterative algorithm for LV.*

INPUT:

(1) Same as Algorithm 7.4.

(2) Sets CLEAR(i) and XUSES(i), $1 \leq i \leq n$, represented by bit vectors of length p, where p is the number variables.

OUTPUT: Sets LVBOT(i), $1 \leq i \leq n$, represented by bit vectors.

METHOD: See procedure LV$ROUND$ROBIN$VERSION in Figure 7.12. □

In Algorithms 7.4 and 7.5, note that we have used rPOSTORDER for the top-down problem and POSTORDER for the bottom-up problem.

procedure AE$ROUND$ROBIN$VERSION
  bit vector NEW, of length m
  proposition CHANGE
  /* Step 1: Initialization. */
  AETOP(1) := 0^m
  for $i := 2$ **to** n **by** 1 **do** AETOP(i) := 1^m **endfor**
  /* Step 2: Iteration. */
  CHANGE := **true**
  while CHANGE **do**
  | CHANGE := **false**
  | **for** $i := 2$ **to** n **by** 1 **do**   /* in rPOSTORDER*/
  | | NEW := **bvand** [(AETOP(k) **bvand** NOTKILL(k)) **bvor** GEN(K)]
  | | $k \in$ PRED(i)
  | | **if** NEW $\neq$ AETOP(i) **then**
  | | AETOP(i) := NEW
  | | CHANGE := **true**
  | | **endif**
  | **endfor**
  endwhile
return

Figure 7.11   Round-robin version of iterative algorithm for AE.

procedure LV$ROUND$ROBIN$VERSION
  bit vector NEW, of length m
  proposition CHANGE
  /* Step 1: Initialization. */
  for $i := 1$ **to** n **by** 1 **do** LVBOT(i) := 0^m **endfor**
  /* Step 2: Iteration. */
  CHANGE := **true**
  while CHANGE **do**
  | CHANGE := **false**
  | **for** $i := n$ **to** 1 **by** -1 **do** /* in POSTORDER */
  | | NEW := **bvor**[(LVBOT(k) **bvand** CLEAR(k)) **bvor** XUSES(k)]
  | | $k \in$ SUC(i)
  | | **if** NEW $\neq$ LVBOT(i) **then**
  | | LVBOT(i) := NEW
  | | CHANGE := **true**
  | | **endif**
  | **endfor**
  endwhile
return

Figure 7.12   Round-robin version of iterative algorithm for LV.

Iterative Algorithm: Round-Robin Version

The proof of termination and correctness of these algorithms is similar in spirit to the proof of Theorem 7.1, so we omit it.

The round-robin version of the iterative algorithm is rather nice theoretically in that its relationship with Algorithm 2.2 is easy to understand, and the analysis of the round-robin version for RFGs is also easy.

Consider the "iterate-until-stabilization" paradigm of Algorithm 2.2. Conceptually we can concatenate the bit vector for LVBOT(1) with that for LVBOT(2) with that for ... with that for LVBOT(n), yielding one bit vector x of size pn. Also, we can construct a function f that essentially visits all nodes in parallel, simultaneously updating LVBOT(i) for each node i. Note that the procedure in Figure 7.12 can be suitably revised along these lines, where x is initially all 0's. We now have Algorithm 2.2, the basic fixed point algorithm of Figure 2.6. The set of bit vectors of length pn forms a lattice, where $x \sqsubseteq y$ iff every bit that is 1 in x is also 1 in y. The zero element of this lattice is the bit vector of all 0's. The length of this lattice is pn, because no chain is longer than pn. Finally, f is monotonic.

Similarly, the "available expressions" problem also fits the mold of the basic fixed point algorithm, Algorithm 2.2. Here we set AETOP(1) to all 0's then ignore it. Bit vectors of size $m(n-1)$ are constructed by concatenating AETOP(2) up to AETOP(n). The zero element of this lattice has all 1's in it, and $x \sqsubseteq y$ iff every bit that is 0 in x is also 0 in y.

We now establish the time complexity of these algorithms for RFGs in terms of bit vector steps and the loop-connectedness parameter of an RFG. Recall that the loop-connectedness of an RFG is the largest number of back arcs on any cycle-free path of that RFG.

LEMMA 7.1 *Any cycle-free path in an RFG G beginning with the initial node is monotonically increasing by rPOSTORDER.*

Proof Any such path must contain no back arcs, and thus is a path in the DAG of G. rPOSTORDER topsorts the DAG of G (Theorem 5.2). □

LEMMA 7.2 *The **while**-loop of Algorithm 7.4 is executed at most $d+2$ times for RFG G, where d is the loop-connectedness of G.*

Proof A 0 propagates from its point of "origin"—a "kill" or the initial node—to the place where it is needed in $d+1$ iterations if it must propagate along a path P of d back arcs. It takes one iteration for a 0 to arrive at the tail of the first back arc of P. This follows since the numbers along the path must be in increasing sequence by Lemma 7.1. After this point, it takes one iteration for a 0 to climb up each back arc in P to the tail of the next back arc, by the same argument. Hence, we need at most $d+1$

iterations to propagate information, plus one more iteration to detect that there are no further changes. □

LEMMA 7.3 *The* **while**-*loop of Algorithm 7.5 is executed at most $d+2$ times for an RFG G.*

Proof A 1 indicating a use propagates backward along a cycle-free path to a given point in $d+1$ iteration if there are d back arcs in the path from the point to the use. It takes one iteration for a 1 to reach the head of the dth back arc in such a path. The proof is then analogous to that of Lemma 7.2. □

THEOREM 7.3 *If we ignore initialization, Algorithm 7.4 (or Algortihm 7.5) takes at most $(d+2)(r+n)$ bit vector steps; that is, $O(dr)$ bit vector steps.*

Proof Since Algorithms 7.4 and 7.5 are almost identical, we mostly refer to Algorithm 7.4 below. We first assume that Algorithm 7.4 has been rewritten so that the expression [(AETOP(k) **bvand** NOTKILL(k) **bvor** GEN(k)] is evaluated for k once on each pass.

For each iteration of the **while**-loop of Algorithm 7.4, $2(n-1)$ bit vector steps are used to evaluate [(AETOP(k) **bvand** NOTKILL(k)) **bvor** GEN(k)] for all k. That is, $n-1$ nodes at 2 bit vector steps per node. (In Algorithm 7.5 this process uses $2n$ bit vector steps.) To perform **bvand** over all predecessors of each node, $r-n$ bit vector steps are aggregately used, because each node with p prodecessors requires $p-1$ bit vector steps, and r predecessors less 1 for each node yields $r-n$. Since the **while**-loop is executed at most $d+2$ times (Lemma 7.2), we have at most $(d+2)(r-n+2n) = (d+2)(r+n)$ bit vector steps. □

COROLLARY 7.2 *In the worst case Algorithms 7.4 and 7.5 require at most $O(r^2)$ bit vector steps, since d can be $O(r)$.*

In practice, Algorithms 7.4 and 7.5 require $O(n)$ bit vector steps. This follows because most flow graphs are sparse ($r = O(n)$), and d is rarely more than 3. Programs written with a disciplined control flow structure have nesting usually no more than 3, and d is essentially the maximum nesting of **while**-loops. Thus, $O(dr)$ becomes $O(n)$ in this situation.

The round-robin version of the iterative algorithm is very easy to understand and program. Irreducible flow graphs are handled without even noticing that they are irreducible. Also, the only preprocessing, computation of rPOSTORDER, can be done in $O(r)$ elementary steps. In contrast, the node listing approach that follows may require some complicated preprocessing. The interval approach, discused in Chapter 8, is not so easy to understand or program.

7.4 Iterative Algorithm: Node Listing Version

The round-robin version requires an extra pass through the nodes to discover that no bits have changed. This and the testing for changed bit vectors on every pass might be avoided if we could somehow know when to halt the iteration. Also, visiting every node on each pass seems unnecessary. The problem is to visit only those nodes whose visitation suffices to propagate information. The node listing approach accomplishes this.

It is evident from the worklist approach and from the proof of Lemma 7.2 that propagation along acyclic paths suffices.

Recall (from Chapter 5) that a strong node listing is a sequence of the nodes of a flow graph that includes every acyclic path as a subsequence. Also, a weak node listing is a sequence of nodes such that every acyclic path P is either a subsequence of the listing or there is another acyclic path which is a subsequence of P.

To solve either "reaching definitions" or "live variables" problems, a weak node listing suffices. However, to solve the "available expressions" problem, a strong node listing is required.

To see that weak node listings suffice for "live variables", consider Figure 5.3 again. Suppose that there is an exposed use of variable X in node 5 and we are trying to determine if X is live at the bottom of node 1. In determining whether there is an X-clear path from node 1 to node 5, we need not consider the path $(1,2,3,4,5)$ because, if this path is X-clear, then so is the path $(1,2,4,5)$.

For the "available expressions" problem, consideration of only basic paths does not suffice. Suppose, in Figure 5.3, that there is a computation of $A+B$ in node 1, and we are trying to determine if $A+B$ is in AETOP(5). Even though the basic path $(1,2,4,5)$ from the bottom of node 1 to the top of node 5 may not contain a recomputation of A or B, A or B may be recomputed in node 3. To see that propagation along simple paths suffices for "available expressions", note that there is no additional information gleaned in propagating a 0, as in Algorithm 7.1, over nonsimple paths.

ALGORITHM 7.6 *Node listing version of iterative algorithm for LV.*

INPUT:
(1) Flow graph $G=(N,A,s)$, $N=\{1,2,\ldots,n\}$, $s=1$.
(2) Weak node listing $L=(x_1,x_2,\ldots,x_g)$ for G.
(3) Sets NOTDEFINED(i) and XUSES(i), $1 \le i \le n$, represented as bit vectors.

OUTPUT: Sets LVBOT(i), $1 \le i \le n$, represented as bit vectors.

METHOD: See procedure LV$NODE$LISTING$VERSION in Figure 7.13. □

procedure LV$NODE$LISTING$VERSION
/* Step 1: Initialize. */
for $i := 1$ **to** n **by** 1 **do** LVBOT(i) := XUSES(i) **endfor**
/* Step 2: Iterate. */
for each node x in L in reverse order **do**
/* Visit node x. */
LVBOT(x) := LVBOT(x) **bvor**
$$\left(\underset{y \in \text{SUC}(x)}{\textbf{bvor}} [\text{NOTDEFINED}(y) \textbf{ bvand } \text{LVBOT}(y)] \right)$$
endfor
return

Figure 7.13 Node listing version of iterative algorithm for LV.

For a bottom-up problem such as LV the nodes in the node listing are visited in reverse order, whereas for a forward problem such as RD the nodes are visited in standard order.

The termination and correctness of Algorithm 7.6 follow immediately from our comments.

THEOREM 7.4 *If each node in G has at most k successors, and node listing L has length g, then Algorithm 7.5 requires at most 2kg bit vector steps.*

Proof Each time a node is visited at most $2k$ bit vector steps are required, and g nodes are visited. □

Exercises

7.1 How can the bit vector problems discussed in Section 7.1 be used to perform the following types of code improvement?
  (a) Constant propagation.
  (b) Dead-code elimination.

7.2 Construct examples of available expressions that are not identified by the model proposed in Section 7.1.1.

7.3 Which data flow analysis problem(s) can be used to detect unitialized uses of local variables in a procedure on
  (a) some path from the initial node, and
  (b) all paths from the initial node?

7.4 For the worklist version of the iterative algorithm, which data structure (stack, queue, or something else) should be used to represent the worklist?

Bibliographic Notes

The "available expressions" problem is discussed in Cocke [1970] and Ullman [1973]. The "reaching definitions" problem is discussed in Allen [1970] and Allen [1971]. The "live variables" problem is discussed in Kennedy [1971], [1975], and [1976]. The "very busy expressions" problem is discussed in Lowry and Medlock [1969], and Schaefer [1973].

Our presentation of the worklist versions of the iterative algorithm comes from Ullman [1973] (integrated version), Kildall [1973] (Kildall's version), and Kou [1975] (segregated version). The round-robin version is discussed in Ullman [1973], Hecht and Ullman [1975], and Kennedy [1976]. Allen and Cocke [1972] employed a round-robin version of an intertive algorithm to find dominators. The node listing variant was discovered by Kennedy [1975]. See the bibliographic notes of Chapter 5 for other references on node listings.

Schaefer [1973], Tenenbaum [1974], and Rosen [1976] also discuss the iterative algorithm for data flow analysis.

There are other algorithms to solve the basic data flow analysis problems treated in this chapter that are asymptotically faster (in the worst case) than the ones we have presented. (In the time bounds that follow, we assume that no node has more than two successors.) Ullman [1973] has discovered an algorithm based on 2–3 trees that used $O(n \log n)$ bit vector steps on RFGs. Graham and Wegman [1975] have a different algorithm for the same class of graphs with the same time bound. However, in those RFGs in which all loops are single-exit or in which the number of exists per loop is bounded, their algorithm takes $O(n)$ bit vector steps. Tarjan [1975] has an $O(n\alpha(n))$ bit vector step algorithm, where $\alpha(n)$ is a very slowly growing function that is related to a functional inverse of Ackermann's function. Kou [1975] has considered parallel models of computation for solving data flow analysis problems.

Chapter 8

INTERVAL ANALYSIS

In this chapter we present an algorithm to solve intraprocedural data flow analysis problems (such as those discussed in Chapter 7) that is commonly referred to as "interval analysis" because it is based on the interval construct (described in Chapter 3).

Our presentation, including the notation, comes from a recent and excellent exposition of interval analysis by F. E. Allen and J. Cocke. There are two important features to this presentation. First, certain information is associated with arcs rather than nodes. Second, an interval ordered, arc listing data structure is used that accommodates a "fast", non-recursive implementation and, furthermore, treats reducible and irreducible flow graphs indistinguishably.

We illustrate the interval analysis algorithm by solving two representative problems: "reaching definitions" and "live definitions". However, most of our exposition is about the former rather than the latter problem. First, we explain how certain data can be associated with arcs, and give the notation we shall employ. Then, an efficient data structure for interval analysis is presented. Next, the algorithm itself is given in three versions: an initial version, a complete but mostly English version, and a definitive Pidgin SIMPL version.

In this chapter we assume WLOG that (a) the initial node of a flow graph G has in-degree 0, and (b) G has exactly one exit node. (Recall that an exit node of a flow graph has out-degree 0.)

Should interval analysis be used in practice? If so, when? Proponents of interval analysis argue that despite its conceptual and programming complexity it is fast in practice, and only anomalous flow graphs (which never occur from real programs) show its true worst-case behavior. Adversaries of interval analysis feel that if you want to go to all the trouble of implementing a complicated algorithm for flow analysis (rather than use an iterative algorithm), you might as well implement one with efficient worst-case behavior. That is, use Ullman's algorithm or the Graham-Wegman algorithm rather than interval analysis. We lean toward the latter side.

Why bother presenting interval analysis here instead of an algorithm whose worst-case time complexity is asymptotically better? There are several reasons. First, average case behavior for typical flow graphs may be more important in practice than worst-case behavior for arbitrary flow

Notation

graphs. Also, the average case behavior of these algorithms has not been well studied. Thus, interval analysis may not be categorically worse than the other algorithms. Second, interval analysis is historically one of the first algorithms for data flow analysis and thus serves as a benchmark for future contenders. Third and finally, interval analysis is representative of algorithms based on reductive graph transformations (such as Ullman's algorithm and the Graham-Wegman algorithm).

8.1 Associating Certain Information with Arcs

To illustrate the interval analysis algorithm, we shall solve both the "reaching definitions" and "live definitions" problems. Our goal is to compute two sets of information for each node in an input flow graph: (1) the set of definitions reaching the top of each node in G, and (2) the set of definitions that are live at the bottom of each node of G. (Recall from Chapter 1 that a definition is live at a point in a flow graph iff that definition can reach that point and can be subsequently used.)

Since a node in a derived graph represents an interval, and an interval may have more than one exit node,[1] different paths through the interval from its header to distinct exit nodes will, in general, define and preserve different sets of data items. In order to represent such possibilities, information defined and preserved by distinct paths through a node will be associated with the corresponding arcs leaving that node. Notice that in the input flow graph G, each node represents just itself. Thus, information associated with the arcs leaving each node x in G can be equivalently associated with the bottom of node x, and vice versa.

Since some information about each node is associated with the arcs leaving that node, it is necessary for the interval analysis algorithm that follows that each node in each graph in the derived sequence have at least one leaving arc. This requirement can be satisfied by adding some pseudo-arcs with no head nodes.

8.2 Notation

In the following, the words 'node' and 'arc' refer to any node or arc in any graph in the derived sequence of an input flow graph G, unless otherwise qualified.

For each node i, let UB_i be the set of locally exposed uses at the top of

---
[1] An exit node of an interval has a successor not in that interval.

node i. For each arc j, let DB_j be the set of locally exposed definitions from the tail node of arc j, and let PB_j be the set of definitions preserved through the tail node of arc j. For the input flow graph G, UB, DB, and PB correspond to XUSES, XDEFS, and PRESERVED.

For each node i, let $RDTOP_i$ be the set of definitions that reach the top of node i, and let $LVTOP_i$ be the set of uses that are upwards exposed (or live) at the top of node i.

For each arc j, let A_j be the set of definitions that are available on (or reach) arc j, and let L_j be the set of definitions that are live on arc j.

We assume that sets such as UB, DB, PB, RDTOP, LVTOP, A, and L are represented by bit vectors, where each bit position corresponds to a definition in the procedure being analyzed. Each use is encoded as a set of definitions so that bit vectors representing definitions and bit vectors representing uses can be meaningfully intersected.

Figure 8.1 contains the basic equations for the "reaching definitions" problem, changed to reflect the use of arcs to hold information. Note that RD7 and RD8 correspond to RD1, and RD9 corresponds to RD2.

RD7. $RDTOP_s$ = OUTSIDEDEFS, where s is the initial node of G, and OUTSIDEDEFS is the set of definitions reaching G from outside.

RD8. $RDTOP_i = \bigcup_p A_p$, for each entering arc p to node i, $i \neq s$.

RD9. $A_j = (RDTOP_i \cap PB_j) \cup DB_j$, for each arc j with tail node i.

Figure 8.1   Equations for "reaching definitions".

Similarly, Figure 8.2 contains the basic equations for the "live variables" problem, namely LV7 and LV8, and the "live definitions" problem, by LD1, changed to reflect arc information. Equations LV7 and LV8 in Figure 8.2 respectively correspond to LV6 of Figure 1.6. The subscript on PB in LV8 refers to arc (i,j).

LV7. $LVTOP_x = UB_x \cup (PRESERVED_x \cap OUTSIDEUSES)$, where x is *the* exit node of G, and OUTSIDEUSES is the set of uses that are upwards exposed (or live) at the bottom of the procedure that G represents. Note that if OUTSIDEUSES = $\emptyset$, then $LVTOP_x = UB_x$.

LV8. $LVTOP_i = \bigcup_{j \in SUC(i)} [LVTOP_j \cap PB_{(i,j)}] \cup UB_i$.

LD1. $L_a = A_a \cap LVTOP_i$, for each arc a with head node i.

Figure 8.2   Equations for "live variables" and "live definitions".

8.3 An Efficient Data Structure for Interval Analysis

In this section we describe an interval ordered, arc listing data structure for the derived sequence of a flow graph. The derived sequence has been altered in two ways. First, each node has at least one leaving arc. Second, arcs leaving an interval from distinct nodes will be represented by distinct arcs in the derived graph.

Example 8.1 Figure 8.3 contains the derived sequence $G = G_1, G_2, G_3$ of a flow graph with initial node 1 and exit node 6. Arcs have been added so that each node has at least one leaving arc. Nodes are numbered consecutively from 1, and arcs are also numbered consecutively from 1 but independent of the nodes. Arc 12 represents arc 3, and arc 13 represents arc 5. Arcs 3 and 5, which are exit arcs from $I(2)$, leave distinct nodes in $I(2)$.

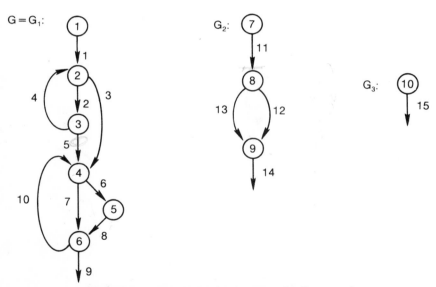

Figure 8.3 The derived sequence of a flow graph.

The data structure for the derived sequence that we shall use is a node ordered arc listing. It consists of five tables named GRAPHS, ARC\$TABLE, FROM\$TO\$TABLE, CORRES\$HEAD, and CORRES\$ARC.

Example 8.2 Figure 8.4 presents these five tables for the derived sequence in Figure 8.3.

GRAPHS

GRAPH #	FIRST$NODE	LAST$NODE	FIRST$ARC	LAST$ARC
1:	1	6	1	10
2:	7	9	11	14
3:	10	10	15	15

ARC$TABLE

NODE	ARC	IN$ARC	OUT$ARC	LATCH	HEAD
1	1		√		√
2	1	√			√
2	4	√		√	
2	2		√		√
2	3		√		√
3	2	√			
3	4		√	√	
3	5		√		
4	3	√			√
4	5	√			√
4	10	√		√	√
4	6		√		√
4	7		√		√
5	6	√			
5	8		√		
6	7	√			
6	8	√			
6	9		√		
6	10			√	
7	11		√		√
8	11	√			
8	12		√		
8	13		√		
9	12	√			
9	13	√			
9	14		√		
10	15		√		√

Figure 8.4 A Representation of the derived sequence in Figure 8.3.

An Efficient Data Structure for Interval Analysis 151

FROMTOTABLE

arc #	FROM	TO		arc #	CORRES$ARC
1:	1	2		1:	—
2:	2	3		2:	—
3:	2	4		3:	—
4:	3	2		4:	—
5:	3	4		5:	—
6:	4	5		6:	—
7:	4	6		7:	—
8:	5	6		8:	—
9:	6	—		9:	—
10:	6	4		10:	—
11:	7	8		11:	1
12:	8	9		12:	3
13:	8	9		13:	5
14:	9	—		14:	9
15:	10	—		15:	14

node #	CORRES$HEAD
1:	—
2:	—
3:	—
4:	—
5:	—
6:	—
7:	1
8:	2
9:	4
10:	7

Figure 8.4 (continued)

For each derived graph there is a row in GRAPHS that gives the FIRST$NODE, LAST$NODE, FIRST$ARC, and LAST$ARC of that derived graph.

ARC$TABLE contains two rows for every arc in the derived sequence (except arcs leaving exit nodes, which have one row): one as an in-arc (entering arc), and one as an out-arc (leaving arc). All in-arcs of a node occur before all out-arcs of that node. Interval order on the nodes is used to order the rows of ARC$TABLE. The six columns in this table are NODE, ARC, IN$ARC, OUT$ARC, LATCH, and HEAD, the last four of which are one-bit indicators.

For each arc there is a row in the FROMTOTABLE that specifies the tail (FROM) and the head (TO) of that arc. For each arc x there is a row in the CORRES$ARC table that specifies the (unique) arc in the next lower derived graph that x represents. For each node i there is a row in the CORRES$HEAD table that gives the head node of the interval that i represents.

8.4 Version 1: An Initial Description

Here is an initial description of interval analysis for "reaching definitions".

Each interval $I(h)$ has only one entry node h. If we know the set RDTOP$_h$ of all definitions that reach the top of node h from *both* inside and outside $I(h)$, then the RDTOP and A sets for every arc and nonheader node in $I(h)$ can be determined from RDTOP$_h$, DB, and PB by processing the nodes in $I(h)$ in interval order.

RDTOP$_h$ is found by a two-phase process. The first phase processes graphs from low order to high order $(G_1, G_2, \ldots, G_m)$, and computes the set of definitions that reach the top of each interval header h from *inside* $I(h)$. The second phase processes graphs from high order to low order $(G_m, G_{m-1}, \ldots, G_1)$, computes the set of definitions that reach the top of each interval header h from *outside* $I(h)$, combines this information with the information from the first phase to obtain RDTOP$_h$, and then uses RDTOP$_h$ accordingly.

During the first phase, for each interval $I(h)$ in each derived graph, DB and PB sets are computed for the arcs leaving the node representing $I(h)$. Also, if $I(h)$ has any latching arcs, RDTOP$_h$ is initially the set union of A_k for each such latch k; otherwise, RDTOP$_h$ is initially $\emptyset$. Thus, the effect of the first phase is to percolate definitions from innermost intervals outward.

To start the second phase, the set of definitions that reach the top of the node representing the limit flow graph G_m from *outside* is OUTSIDE-DEFS. This is the set of definitions that also reaches the top of the initial node of G_{m-1} from outside. Now all reaching definitions in G_{m-1} can be computed. Then, since each node in G_{m-1} represents an interval in G_{m-2}, the definitions that reach interval headers in G_{m-2} from outside the intervals can be gleaned from the just computed G_{m-1} information. Eventually, information is propagated from G_m back to G_1.

The algorithm for computing "live definitions" is analogous to the above, and it too requires two phases. The first phase, which computes information from inside each interval, can be imbedded in the first phase of the "reaching definitions" algorithm. The second phase processes the nodes of each interval in reverse interval order, so it cannot be combined with the second phase for "reaching definitions".

8.5 Version 2: A More Detailed Description

Here is a more detailed description of the interval analysis algorithm for "reaching definitions".

ALGORITHM 8.1 *Interval analysis for "reaching definitions" problem.*
INPUTS:
(1) The derived sequence $\bar{G} = G_1, G_2, \ldots, G_m$.
(2) The intervals in each graph with their nodes in interval order.
(3) The DB and PB sets for each arc in G.
(4) The set OUTSIDEDEFS.
OUTPUTS:
(1) For each node i, the set $RDTOP_i$ of definitions that reach the top of node i.
(2) For each arc j, the set A_j of definitions available on arc j.
METHOD: The algorithm first calls procedure PHASE1 in Figure 8.5, then calls procedure PHASE2 in Figure 8.6. □

---

procedure PHASE1 /* Mostly English Version */
1. For each graph G_g, in the order $G_1, G_2, \ldots, G_{m-1}$, perform steps 2 and 3.
2. If the current graph is not G_1, then initialize the PB and DB sets of G_g as follows. First identify (interval exit) arcs in G_{g-1} to which each arc in G_g corresponds. Then, use the P and D information generated during step 3 for G_{g-1} as follows. For each arc i in G_g with corresponding exit arc x from interval with head h in G_{g-1}, do
   2.1. $PB_i := P_x$, and
   2.2. $DB_i := (RDTOP_h \cap P_x) \cup D_x$.
3. For each exit arc of each interval in G_g determine P, the definitions preserved on some path through the interval to the exit, and D, the definitions in the interval that may be available on exit. These are determined by finding P and D for each arc in the interval as follows.
   3.1. For each exit arc i of the header node,
      $P_i := PB_i$, and
      $D_i := DB_i$.
   3.2. For each exit arc i of each nonheader node j in interval order,
      $P_i := (\bigcup_p P_p) \cap PB_i$, and
      $D_i := [(\bigcup_p D_p) \cap PB_i] \cup DB_i$, for all input arcs p to node j.
   While processing an interval, determine the set $RDTOP_h$ of definitions that can reach the interval head h from inside the interval by
      $RDTOP_h := \bigcup_k D_k$, for all latches k of $I(h)$ that enter $I(h)$.
   If $I(h)$ has no latches, then $RDTOP_h := \emptyset$.
return

Figure 8.5 Mostly English version of PHASE1.

procedure PHASE2 /* Mostly English Version */
1. The RDTOP set for the single node that represents G_m is initialized to OUTSIDEDEFS, the set of definitions know to reach the procedure under analysis from outside.
2. For each graph G_g, in the order $G_{m-1},\ldots,G_2,G_1$, perform steps 3 and 4.
3. For each node i in G_{g+1},

    $\text{RDTOP}_h := \text{RDTOP}_h \cup \text{RDTOP}_i$,

    where h is the head of the interval in G_g, which i represents in G_{g+1}.
4. For each interval process the nodes in the interval order determining the definitions reaching each node and available on each node exit arc as follows:
    4.1. For each exit arc i of the header node h,

    $A_i := (\text{RDTOP}_h \cap \text{PB}_i) \cup \text{DB}_i$.

    4.2. For each nonheader node j in the interval order, first form

    $\text{RDTOP}_j := \bigcup_p A_p$, for all arcs p entering node j,

    then for each exit arc i of node j form
    $A_i = (\text{RDTOP}_j \cap \text{PB}_i) \cap \text{DB}_i$.

return

Figure 8.6 Mostly English version of PHASE2

```
/* Global declarations in accompanying text go here. */
integers G, I   /* used as loop indices */
procedure INTERVAL$ANALYSIS
   call PHASE1   /* Process graphs from low order to high order. */
   call PHASE2   /* Process graphs from high order to low order. */
   call PHASE3   /* Process graphs from high order to low order. */
return
procedure PHASE1
   /* See Figure 8.8. */
return
procedure PHASE2
   /* See Figure 8.9. */
return
procedure PHASE3
   /* See Figure 8.10. */
return
start INTERVAL$ANALYSIS   /* Begin execution with this procedure. */
```

Figure 8.7 Interval analysis algorithm for "reaching definitions" and "live definitions".

8.6 Version 3: A Pidgin SIMPL Description

Here is a definitive Pidgin SIMPL version of the interval analysis algorithm for solving both the "reaching definitions" and "live definitions" problems.

ALGORITHM 8.2 *Interval analysis for "reaching definitions" and "live definitions" problems.*

INPUTS:

(1) The derived sequence $G = G_1, G_2, \ldots, G_m$ of a flow graph G represented by the five tables GRAPHS, ARC$TABLE, FROM$TO$TABLE, CORRES$HEAD, and CORRES$ARC.
(2) The DB and PB sets for each arc in G, and the UB set for each node in G.
(3) OUTSIDEDEFS and OUTSIDEUSES.

OUTPUTS:

(1) $RDTOP_i$, for each node i.
(2) $LVTOP_i$, for each node i.
(3) L_j, for each arc j.

METHOD: See Figure 8.7. □

8.7 Comments

For irreducible flow graphs, certain nodes in the derived sequence must be split (as described in Chapter 6) and the tables of the node ordered, arc listing data structure must be altered appropriately. However, once the node ordered, arc listing data structure is so altered, the interval analysis algorithm in the previous section treats RFGs and IRFGs indistinguishably.

Example 8.3 Each node in G_1 of Figure 8.11 is an interval head. G_2 is G_1 with node 3 split. Nodes 5 and 7 are really copies of node 3. Thus, node 3 is the corresponding head of nodes 5 and 7, and arc 4 is the corresponding arc of arcs 8 and 10.

As for timing, there are two parts to the interval analysis algorithm: constructing the data structure from the input flow graph, and propagating information. To construct the interval ordered, arc listing data structure, Algorithm 3.2 can be used on each graph in the derived sequence to construct the next graph. Each execution of Algorithm 3.2 requires time proportional to the number of arcs in the input graph. In the very worst

```
procedure PHASE1
  integer HEAD
  for G := 1 to NO$OF$GRAPHS − 1 by 1 do
    if G ≠ 1 then
      /* Pick up (arc and node) data from lower graph. */
      for I := FIRST$ARC(G) to LAST$ARC(G) by 1 do
        PB(I) := P(CORRES$ARC(I))
        DB(I) := [RDTOP(CORRES$HEAD(FROM(I))) bvand PB(I)]
                  bvor D(CORRES$ARC(I))
      endfor   /* I */
      for I := FIRST$NODE(G) to LAST$NODE(G) by 1 do
        UB(I) := U(CORRES$HEAD(I))
        RDTOP(I) := RDTOP(CORRES$HEAD(I))
      endfor   /* I */
    endif
    HEAD := 0
    for I := 2*FIRST$ARC(G) − 1 to 2*LAST$ARC(G) by 1 do
      if HEADER(I) then
        if HEAD = NODE(I) then HEAD := NODE(I) endif
        if OUT$ARC(I) then
          P(ARC(I)) := PB(ARC(I))
          D(ARC(I)) := DB(ARC(I))
        endif
      else
        if IN$ARC(I) then
          PIN(NODE(I)) := PIN(NODE(I)) bvor P(ARC(I))
          DIN(NODE(I)) := DIN(NODE(I)) bvor D(ARC(I))
          LVTOP(HEAD) := LVTOP(HEAD) bvor [P(ARC(I)) bvand
                                          UB(NODE(I))]
        else
          P(ARC(I)) := PIN(NODE(I)) bvand PB(ARC(I))
          D(ARC(I)) := [DIN(NODE(I)) bvand PB(ARC(I))]
                        bvor DB(ARC(I))
        endif
      endif
      if LATCH(I) then RDTOP(HEAD) := RDTOP(HEAD) bvor D(ARC(I))
      endif
    endfor   /* I */
  endfor   /* G */
return
```

Figure 8.8   PHASE1 of interval analysis.

```
procedure PHASE2
   RDTOP(NO$OF$NODES) := OUTSIDEDEFS
   for G := NO$OF$GRAPHS − 1 to 1 by − 1 do
      for I := FIRST$NODE(G + 1) to LAST$NODE(G + 1) by 1 do
         RDTOP(CORRES$HEAD(I)) := RDTOP(CORRES$HEAD(I))  bvor
                                                         RDTOP(I)
      endfor /* I */
      for I := 2*FIRST$ARC(G) − 1 to 2*LAST$ARC(G) by 1 do
         if IN$ARC(I) ∧ ⌐ HEADER(I) then
            RDTOP(NODE(I)) := RDTOP(NODE(I)) bvor A(ARC(I))
         endif
         if OUT$ARC(I) then
            A(ARC(I)) := [RDTOP(NODE(I)) bvand PB(ARC(I))] bvor
                         DB(ARC(I))
         endif
      endfor /* I */
   endfor /* G */
return
```

Figure 8.9 PHASE2 of interval analysis.

```
procedure PHASE3
   LVTOP(NO$OF$NODES) :=
LVTOP(CORRES$HEAD(NO$OF$NODES))
   UB(NO$OF$NODES) := LVTOP (CORRES$HEAD(NO$OF$NODES))

   for G := NO$OF$GRAPHS − 1 to 1 by − 1 do
      for I := FIRST$NODE(G + 1) to LAST$NODE(G + 1) by 1 do
         LVTOP(CORRES$HEAD(I)) := LVTOP(CORRES$HEAD(I)) bvor UB(I)
      endfor /* I */
      for I := 2*LAST$ARC(G) to 2*FIRST$ARC(G) − 1 by − 1 do
         if OUT$ARC(I) ∧ ⌐ HEADER(I) then
            LVTOP (NODE(I)) := LVTOP (NODE(I)) bvor [LVTOP(TO(ARC(I)))
                               bvand PB(ARC(I))]
         endif
         if IN$ARC(I) then
            L(ARC(I)) := LVTOP (NODE(I)) bvand A(ARC(I))
         endif
      endfor /* I */
   endfor /* G */
return
```

Figure 8.10 PHASE3 interval analysis.

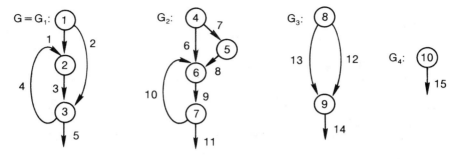

Figure 8.11 An irreducible flow graph.

case, if there are k graphs in the derived sequence, then we need

$$\sum_{i=0}^{k-1}(r-i) = kr - \frac{k(k-1)}{2}$$

or $O(kr)$ elementary steps if $r \gg k$. There are some anomalous flow graphs with $k = O(r^2)$. Also, irreducible flow graphs present an additional time and programming complication. So, just constructing the data structure alone can be very slow in the worst case. In practice, though, most flow graphs are found to be reducible with k rarely above 6 for real programs. Thus, the time to construct the required data structure for real programs may not be bad at all.

As for the second part of the algorithm that propagates information, let n' be the number of nodes and let r' be the number of arcs in the derived sequence. The number of bit vector steps for interval analysis is easily seen to be $O(\max(n',r'))$. Since flow graphs are connected and each node has at least one leaving arc, $\max(n',r') = r'$. So, propagation requires $O(r')$ bit vector steps. Although there are some degenerate flow graphs with $O(r^2)$ arcs in their derived sequence, we conjecture that it is typically the case that $O(r') = O(r)$ for flow graphs of real programs.

Bibliographic Notes

The exposition of interval analysis in this chapter comes from Allen and Cocke [1976].

Interval analysis was developed by Cocke [1970], and Allen [1970]. The interval concept is due to Cocke. Other references on interval analysis include Cocke and Schwartz [1970], Allen [1971], Kennedy [1971], Aho and Ullman [1973], Schaefer [1973], and Kennedy [1976]. Kennedy [1971] showed how the original method of Cocke and Allen could be used to solve bottom-up problems such as "live variables". Knuth [1971] describes an empirical study of FORTRAN programs that provides some important observations for code improvement.

Chapter 9

MONOTONE DATA FLOW ANALYSIS FRAMEWORKS

Continued research on code improvement and binding time analysis has uncovered some data flow analysis problems that do not fit the mold of the bit vector problems described in Sections 1.5.1 and 7.1. For example, Kildall [1973] studied the problems of constant propagation and common subexpression elimination. He represented information for the former problem by sets of pairs of variables and constants, and information for the latter problem by equivalence relations. Kildall couched these problems in a semilattice-theoretic "framework" (which also models the bit vector problems we have seen already), and proposed an iterative algorithm for their solution. Other data flow analysis problems that do not naturally fit the mold of the bit vector problems include the type determination problem of Tenenbaum [1974], the automatic data structure selection problem studied by Schwartz [1975], and induction variable analysis of Fong, Kam, and Ullman [1975].

In this chapter we consider an abstraction for a data flow analysis problem called a "monotone framework". One important subclass of the monotone frameworks is called the "distributive frameworks". Both of these terms will be defined in Section 9.2. There are several important solutions to a monotone framework: "the meet over all paths (MOP) solution" defined in Section 9.3, and "the maximum fixed point (MFP) solution" to certain sets of simultaneous equations defined in Section 9.5.

Kildall made some very important observations and proved some key results about data flow analysis problems. He observed that in a data flow analysis problem, the relevant information can be nicely modeled by a semilattice with a zero element and of finite length,[1] and, the effect of information due to a basic block can be viewed as an operation on the semilattice. The desired solution of a data flow analysis problem is the MOP solution. This solution can be interpreted informally as the calculation for each node in the program flow graph of the maximum information, relevant to the data flow analysis problem, which can be derived from

---

[1] We shall review in this chapter the definitions of terms such as 'semilattice', 'of finite length', and so forth, which first appeared in Section 2.3 and have been ignored up to now.

Justification for a More General Setting

every possible execution path from the initial node of the flow graph to that particular node.

Kildall studied distributive frameworks. He proved that (a worklist version of) the iterative algorithm converges to the MFP solution of a framework, independent of the order in which nodes are visited. Also, for a distributive framework the MFP solution is equal to the MOP solution.

Graham and Wegman [1976] and Fong, Kam, and Ullman [1975] have studied "fast frameworks", a restricted subclass of distributive frameworks in which the MOP solution can be found efficiently. Although we define such frameworks in this chapter, we have not included those algorithms for fast frameworks contained in the above two references.

Kam and Ullman [1975] have considered a generalization of Kildall's semilattice model that they call a monotone[1] data flow analysis framework (or monotone framework for short), which appears to be the most appropriate formal model for data flow analysis problems found so far. They proved that when Kildall's iterative algorithm is applied to a monotone framework, it obtains the MFP solution to a certain set of simultaneous equations. Also, the MOP problem for a monotone framework is undecidable.

In this chapter we just summarize some of the known results about monotone frameworks. (Most proofs are omitted, not because of their general difficulty, but because of their aggregate length.) Our treatment begins with justification for a more general setting. Both monotone frameworks and distributive frameworks are defined next, and illustrated with two running examples: the bit vector version of the available expressions problem, and the constant propagation problem. The meet over all paths solution is defined for monotone frameworks, and is subsequently shown to be undecidable in general and for a particular framework. Finally, the maximum fixed point solution of a framework is defined, and some versions of the general iterative algorithm are presented.

This chapter is important for both the practitioner and people with an appreciation for the theoretical foundation of data flow analysis. Section 9.5 contains useful algorithms for the practitioner, although it will be very difficult to understand them without reading Sections 9.1 and 9.2 first.

9.1 Justification for a More General Setting

Why is the bit vector model inappropriate in general for data flow analysis problems? One reason for this is that often there is no natural way to represent the information to be propagated as bits, where propagation proceeds from bit vector AND and OR operations. Another reason is that sometimes the iterative algorithm for a data flow analysis problem either does not halt (i.e., the information propagating does not stabilize) or else more than $d+2$ (round-robin) iterations are necessary to propagate the information.[2]

Kildall [1973] presents two examples of data flow analysis problems where the bit vector paradigm is inappropriate: constant propagation, and common subexpression elimination with equivalence relations.

Suppose that we desire to perform *constant propagation* (AKA *folding*). For this problem we seek to discover, at the top (or bottom) of each flow graph node, those variables that are assigned the same constant value on all paths from the initial node to the top (or bottom) of the given node. One way to represent information for constant propagation at a node is by a finite subset of $V \times C$, where V is the set of program variables and C is the set of possible values of constants. For instance, we can represent the fact that variables A and B are assigned the values 1 and 2 respectively on all paths from the initial node to the top of a given node x by associating $\{(A,1),(B,2)\}$ with the top of node x. To derive such information we can use the operation of set intersection to advantage. However, a bit vector representation of finite subsets of $V \times C$ is inappropriate, in general, if there is not a small finite bound on $\#C$.

Kildall has proposed the use of an equivalence relation to help detect available expressions and to remove redundant assignments. The idea is to determine at the top of each flow graph node an equivalence relation (containing variables and expressions in its equivalence classes) that not only shows what expressions are available at the top of that node but what variables always hold equivalent values there. The iterative algorithm can be used to compute these equivalence relations. Equivalences implied by assignment statements within a node affect the equivalence relations, and "refinement" is used to "unify" two equivalence relations at the top of a node instead of the bit vector AND operation. Figure 9.1 shows an example where the equivalence relation approach detects an available expression that the bit vector approach misses.

A well-known example of a data flow analysis problem in which the iterative algorithm may not halt is type determination. Suppose we are propagating possible types of each variable. The situation shown in Figure 9.2 gives X the infinite set of types "integer, or set of integers, or set of set of integers, or...".[3] The usual approach in this kind of problem is to limit the "set of" operation to a depth of 2 or 3, and anything more complex is labeled "don't know". Otherwise, the iterative algorithm discovers a new type for X on each iteration.

[2] Recall that d is the depth of a flow graph. See Lemma 7.2.
[3] Fong, Kam, and Ullman [1975] show how to detect such self-dependence of types.

Definition of a Monotone Framework

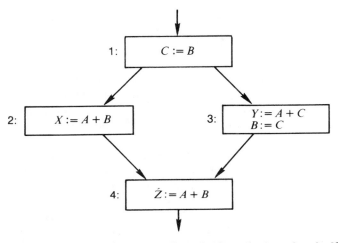

Figure 9.1 Is the value of $A + B$ available at the top of node 4?

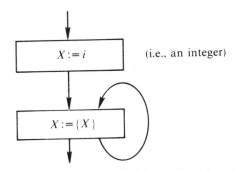

Figure 9.2 Self-dependence of type information.

An example of an instance of a data flow analysis problem requiring at least $2d+2$ (round-robin) iterations of the iterative algorithm before it halts can be found in Kam and Ullman [1976].

Certainly, a more general setting than the bit vector model is necessary.

9.2 Definition of a Monotone Framework

In this section we define the abstract notion of a monotone framework, and illustrate it with two running examples: the (bit vector) available expressions problem, and the constant propagation problem.

Recall that a semilattice is a pair $(L, \sqcap)$, where L is a nonempty set, and $\sqcap$ is a binary operation on L (called *meet*), which is idempotent, commutative, and associative. We assume that L has a zero element **0**, but not necessarily a one element **1**. In addition, we assume that L is of finite length: each strictly decreasing chain $x_1 \sqsupset x_2 \sqsupset \cdots$ of elements of L is finite.

Semilattice elements represent information that can be posted at the top of a node of a flow graph. The meet operation represents the effect of information converging from paths. The effect of the contents of each node on a semilattice element is modeled by an operation on L.

Example 9.1 The available expressions problem (as defined in Section 7.1.1) is easily fit into this setting. Let $(L, \sqcap)$ be the semilattice where L is the set of 2^m m-bit vectors. Here m is any arbitrary fixed positive integer representing the number of distinct expressions in a flow graph. The meet operation of two vectors in L is the logical bit vector AND. The meaning of $x \sqsubseteq y$ in this context is that bit vector y has a 1 in whatever positions x has 1, and perhaps some others. The vector of all **0**s is the zero element **0** of L. The vector of all 1's is the one element **1** of L. If a node has KILL and GEN of k and g respectively, then the function of that node can be represented by $\langle k, g \rangle$, where $(\forall x \in L)[\langle k, g \rangle(x) = (x \text{ AND NOT } k) \text{ OR } g]$.

Example 9.2 The constant propagation problem may also be cast in this semilattice mold. Let $(L, \sqcap)$ be a semilattice in which $L \subset 2^{V \times R}$, where V is an infinite set of variables, and R is the set of real numbers. L is the set of functions from finite subsets of V to R. The zero element of L is the empty set. L has no one element. The meet operation is set intersection (i.e., intersection of sets of ordered pairs).

Intuitively, $x \in L$ represents the information about variables that we may assume at certain points of a program flow graph. $(A, r) \in x$ means that variable A has value r. That is, each element of L is a set of ordered pairs, where the first component is a variable, and the second component is a value. Clearly, we want each element of L to be a function: if a variable has more than one value, it is not a candidate for folding.

Each flow graph node may be modeled by an operation on L defined as follows. For simplicity, let us assume that flow graph nodes contain only assignment statements with form $A := B \theta C$ and $A := r$, where $A, B, C \in V$, $r \in R$, and $\theta \in \{+, -, *, /\}$. Furthermore, we ignore the aliasing problem here. The function representing a node consists of the composition of the functions representing statements within a node. If $x \in L$, then

(i) $\langle A := r \rangle(x) = y$, where y is formed from x by adding pair (A, r) to x and deleting any other pair with first component A.

Definition of a Monotone Framework

(ii) $\langle A := B\theta C \rangle(x) = y$, where, if (B,b) and (C,c) are in x, then we do as in case (i) but use $b\theta c$ for r.

DEFINITION Given a semilattice $(L, \sqcap)$ of finite length with a zero element, a set of operations F on L is said to be a *monotone operation space associated with L* iff the following four conditions are satisfied:

(C1) Each $f \in F$ is monotonic. That is, $(\forall f \in F)(\forall x, y \in L)$ $[x \sqsubseteq y \rightarrow f(x) \sqsubseteq f(y)]$. By Observation 2.17, this is equivalent to the GKUW property
$(\forall f \in F)(\forall x, y \in L)[f(x \sqcap y) \sqsubseteq f(x) \sqcap f(y)]$.

(C2) There exists an identity operation e in F. That is, $(\exists e \in F)(\forall x \in L)[e(x) = x]$.

(C3) F is closed under composition. That is, $(\forall f, g \in F)[fg \in F]$, where $(\forall x \in L)[fg(x) = f(g(x))]$.

(C4) For each $x \in L$, there exists an $f \in F$ such that $x = f(\mathbf{0})$.

DEFINITION A *monotone data flow analysis framework* (or *monotone framework*) is a triple $D = (L, \sqcap, F)$, where

(1) $(L, \sqcap)$ is a semilattice of finite length with a zero element $\mathbf{0}$, and
(2) F is a monotone operation space associated with L.

What is the justification for the four conditions defining F? Intuitively, they help allow us to use Algorithm 2.2. Condition C1 is observed in virtually all problems that are called data flow analysis problems in the literature on programming languages. Condition C2 comes from the assumption that a node may contain no statements that directly transform information as control flows through it. Condition C3 reflects the action of passing information through two successive nodes. Condition C4 prevents the inclusion of irrelevant information in L, although a weaker C4 may be appropriate.

DEFINITION An *instance* of a monotone framework is a pair $I = (G, M)$, where

(1) $G = (N, A, s)$ is a flow graph, and
(2) $M : N \rightarrow F$ is a function that maps each node in N to an operation in F.

One important subclass of monotone frameworks is the "distributive frameworks".

DEFINITION A montone framework $D = (L, \sqcap, F)$ is called a *distributive* *framework* iff

(C5) $(\forall f \in F)(\forall x, y \in L)[f(x \sqcap y) = f(x) \sqcap f(y)]$.

Condition C5 is called distributivity here and a meet-endomorphism in Section 2.3.5. Recall from Observation 2.19 that C5 implies C1.

Example 9.3 Let AVAIL $=(L, \sqcap, F)$, where $(L, \sqcap)$ is defined as in Example 9.1, and F is the set of 2^{m+1} operations on L defined by $F = \{\langle k, g \rangle \mid k$ and g are each m-bit vectors and $(\forall x \in L)[\langle k, g \rangle(x) = (x \text{ AND NOT } k) \text{ OR } g]\}$. AVAIL is a distributive (and thus monotone) framework by Theorem 9.1 below.

THEOREM 9.1 *AVAIL $=(L, \sqcap, F)$ as defined in Example 9.3 is a distributive framework.*

Proof It is easy to see that AVAIL is a semilattice with a zero element, and of finite length.

We must now verify that F is a distributive operation space associated with L.

To show distributivity, we must show that $(\forall x, y \in L)(\forall f \in F)[f(x \sqcap y) = f(x) \sqcap f(y)]$. Let $x, y \in L$, and let $f = \langle k, g \rangle \in F$. It suffices to show that, for all i where $1 \leq i \leq m$, the ith bit of $f(x \sqcap y)$ equals the ith bit of $f(x) \sqcap f(y)$.

(i) Suppose the ith bit of $x \sqcap y$ is 0. Then either the ith bit of x or y is 0. WLOG, assume the ith bit of x is 0. Then the ith bit of $f(x)$ equals $((0 \text{ AND } (\text{NOT } i\text{th bit of } k)) \text{ OR } i\text{th bit of } g)$, which equals the bit of g, which equals the ith bit of $f(x \sqcap y)$.

(ii) Suppose the ith bit of $x \sqcap y$ is 1. Then (ith bit of x) = (ith bit of y) = 1. It follows that (ith bit of $f(x \sqcap y)$) = (ith bit of $f(x)$) = (ith bit of $f(y)$) = (AND (NOT ith bit k)) OR (ith bit of g).

The identity operation on F is $\langle 0, 0 \rangle$, because $((x \text{ AND NOT } 0) \text{ OR } 0) = x \text{ AND } 1 = x$, for all $x \in L$.

To see that F is closed under composition, let $\langle k1, g1 \rangle$ and $\langle k2, g2 \rangle$ be any two operations in F. It is a simple and delightful exercise in Boolean algebra to show that $(\forall x \in L)[\langle k2, g2 \rangle(\langle k1, g1 \rangle(x)) = \langle k3, g3 \rangle(x)]$, where $k3 = (k1 \text{ AND NOT } g1) \text{ OR } k2$, and $g3 = (g1 \text{ AND NOT } k2) \text{ OR } g2$.

Finally, condition C4 of the definition of a framework is easily satisfied, since F contains all possibilities. □

Definition of a Monotone Framework

Example 9.4 Let $CONST = (L, \sqcap, F)$ represent the constant propagation problem, where $(L, \sqcap)$ is as defined in Example 9.2. Let F include operations denoted by $\langle A := B\theta C \rangle$ and $\langle A := r \rangle$ for all A, B, C, r, and θ as specified in Example 9.2. In addition we assume that F contains an identity operation $\mathbf{e}$, and that if f and g are in F then so is fg. CONST is a monotone (but not distributive) framework as we shall see in Theorem 9.2.

The following lemma is used in the proof of Theorem 9.2.

LEMMA 9.1 *Let L be a semilattice, and let $f_1, \ldots, f_n$ be operations on L. If each f_i is monotonic, then*

$$(\forall x, y \in L)\left[f_1 f_2 \cdots f_n(x \sqcap y) \sqsubseteq f_1 f_2 \cdots f_n(x) \sqcap f_1 f_2 \cdots f_n(y) \right].$$

Proof We use backward induction on i.
BASIS: $(i = n)$. $f_n(x \sqcap y) \sqsubseteq f_n(x) \sqcap f_n(y)$ by assumption.
INDUCTIVE STEP: $(i < n)$. Assume that $f_i \cdots f_n(x \sqcap y) \sqsubseteq f_i \cdots f_n(x) \sqcap f_i \cdots f_n(y)$. Then $f_{i-1}(f_i \cdots f_n(x \sqcap y)) \sqsubseteq f_{i-1}(f_i \cdots f_n(x) \sqcap f_i \cdots f_n(y)) \sqsubseteq f_{i-1} \cdots f_n(x) \sqcap f_{i-1} \cdots f_n(y)$, both by assumption. □

THEOREM 9.2 *$CONST = (L, \sqcap, F)$ as defined in Example 9.4 is a monotone framework, but is not distributive.*

Proof It is easy to see that L is a semilattice of finite length with a zero element. Furthermore, for any element $x \in L$, $x = f_1 f_2 \cdots f_m(\mathbf{0})$, for some integer m, where each f_i is of the form $\langle A := r \rangle$. So to show that F is a monotonic operation space associated with L, it suffices by Lemma 9.1, to show that for all $x, y \in L$ and all operations of the form $\langle A := B\theta C \rangle$ or $\langle A := r \rangle$, $\langle A := B\theta C \rangle(x \sqcap y) \sqsubseteq \langle A\theta C \rangle(x) \sqcap \langle A := B\theta C \rangle(y)$ and $\langle A := r \rangle(x \sqcap y) \sqsubseteq \langle A := r \rangle(x) \sqcap \langle A := r \rangle(y)$.

Recall that $\sqcap$ is set intersection, and $\sqsubseteq$ is set inclusion in CONST.

(i) Let $\langle A := B\theta C \rangle \in F$, and let $x, y \in L$. Consider $z = \langle A := B\theta C \rangle(x \sqcap y)$. For all $X \in V - \{A\}$, if $(X, r) \in z$ then $(X, r) \in x$ and $(X, r) \in y$. Hence, $(X, r) \in \langle A := B\theta C \rangle(x)$ and $(X, r) \in \langle A := B\theta C \rangle(y)$. If A is undefined in z, then we are done. Suppose however, that $(A, r) \in z$. Then $\{(B, r_1), (C, r_2)\}$ is a subset of both x and y, for some r_1 and r_2 such that $r = r_1 \theta r_2$. This implies that $(A, r) \in \langle A := B\theta C \rangle(x)$ and $(A, r) \in \langle A := B\theta C \rangle(y)$. Thus, $\langle A := B\theta C \rangle$ is monotonic.

(ii) The argument that $\langle A := r \rangle$ is montonic is very easy.

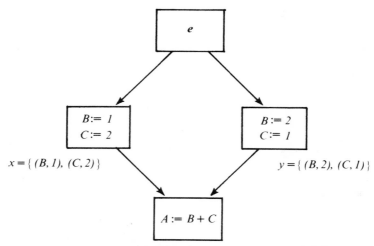

Figure 9.3 Counter example to distributivity of CONST.

To show that CONST is not distributive, consider the flow graph in Figure 9.3. There, $\langle A := B + C \rangle(x \sqcap y) = \emptyset$, while $\langle A := B + C \rangle(x) \sqcap \langle A := B + C \rangle(y) = \{(A, 3)\}$. □

A "fast framework" is a special case of a distributive framework. Fong, Kam, and Ullman [1975] and Graham and Wegman [1976] present efficient algorithms (based on reductive flow graph transformation) for fast frameworks.

NOTATION Let f, g be two distributive operations on a semilattice L. Then $f \sqcap g$ stands for the operation on L such that $(\forall x \in L)[f \sqcap g(x) = f(x) \sqcap g(x)]$.

For any distributive operation f on L, we let $f^0 = \mathbf{e}$, the identity operation, and f^j represent the j-fold composition of f.

We let f^* represent $f^0 \sqcap f^1 \sqcap f^2 \sqcap \cdots$.

DEFINITION A distributive framework $D = (L, \sqcap, F)$ is called a *fast framework* iff it satisfies the following:

(C6) $(\forall f, g \in F)[f \sqcap g \in F]$.

(C7) $(\forall f \in F)(\exists k) \left[\bigsqcap_{i=0}^{k} f^i = f^* \right]$.

THEOREM 9.3 AVAIL *is a fast framework*.

The Meet Over All Paths (MOP) Solution

Proof We need to verify conditions C6 and C7. For C6, let $\langle k1,g1 \rangle$ and $\langle k2,g2 \rangle$ be any two operations in F. Then, $\langle k1,g1 \rangle \sqcap \langle k2,g2 \rangle = \langle (k1 \text{ AND } k2) \text{ OR } (k1 \text{ AND NOT } g1) \text{ OR } (k2 \text{ AND NOT } g2), g1 \text{ OR } g2 \rangle$ is in F.

Satisfaction of C7 by AVAIL is left as an exercise to the reader. □

9.3 The Meet Over All Paths (MOP) Solution

It appears generally true that for data flow analysis problem, we search for the "meet over all paths" solution. Intuitively, this solution is the calculation for each node in the program flow graph of the maximum information, relevant to the problem at hand, which can be derived from every possible execution path from the initial node to that node.

Example 9.5 Suppose that we seek to determine the set of all variables that are assigned the same constant value on *all* paths from the initial node in Figure 9.4 to the the top of each node there. The intuitive way to compute this information is to compute for each path P from 1 to i the set of pairs (V,c) such that variable V has constant value c if computation follows path P from 1 to i. The desired answer is an intersection over all such paths of the sets of obtained pairs. In this case we can compute the intersection of the information over all such paths, even though there are an infinite number of paths from 1 to 2 and from 1 to 3.

NOTATION Given an instance $I=(G,M)$ of a montone framework $D=(L,\sqcap,F)$, we let f_i denote $M(i)$, the operation F associated with node i.

Let $Q = x_1, x_2, \ldots, x_{m-1}, x_m, x_{m+1}$ be a path in G. We shall use $f_Q(\cdot)$ for

$$f_{x_m}\left(f_{x_{m-1}}\left(\cdots \left(f_{x_1}(\cdot)\right)\cdots\right)\right).$$

Note that $f_{x_{m+1}}$ is not in the composition.

DEFINITION Given a montone framework $D=(L,\sqcap,F)$ and an instance $I=(G,M)$ of D, where $G=(N,A,s)$, the *meet over all paths solution (MOP)* for I is the following assignment of information with the top of each node in G:

$$A[s] = \mathbf{0}$$

$$A[x] = \bigsqcap_{Q \in \text{PATH}(x)} f_Q(\mathbf{0}),$$

for all $x \in N - \{s\}$, where $\text{PATH}(x) = \{Q | Q \text{ is a parh in } G \text{ from } s \text{ to } x\}$. (Note that $\sqcap$ converges by the definition of a monotone framework.)

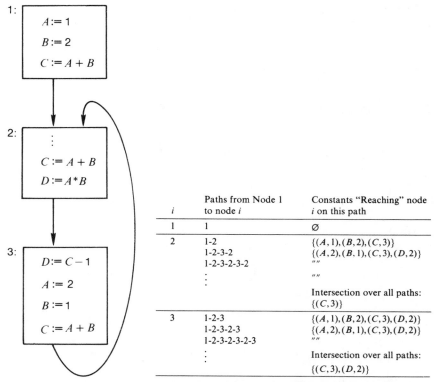

Figure 9.4 An illustration of computing the MOP solution for an instance of CONST.

9.4 Undecidability of the MOP Problem for Monotone Frameworks

In this section we show that there does not exist an algorithm that, for an instance $I = (G, M)$ of an arbitrary monotone framework $D = (L, \sqcap, F)$, will compute the MOP solution. Furthermore, the MOP problem is undecidable for CONST.

DEFINITION The *modified post's correspondence problem* (MPCP) is described as follows. Given arbitrary lists A and B of k strings each, where $A = a_1, a_2, \ldots, a_k$ and $B = b_1, b_2, \ldots, b_k$ with each a_i and b_i in $\{0,1\}^+$,[4] we have $a_1 a_{i_1} a_{i_2} \cdots a_{i_r} = b_1 b_{i_1} b_{i_2} \cdots b_{i_r}$.

---

[4] $\{0,1\}^+$ denotes the set of all finite length strings of 0's and 1's of length at least one.

It is well known that MPCP is undecidable.[5]

Given an instance AB of MPCP, we can construct a monotone framework $D_{AB} = (L_{AB}, \sqcap, F_{AB})$ as follows.

The elements of L_{AB} are

(1) **0**, the semilattice zero,
(2) the special element $, which will in a sense indicate nonsolution to MPCP, and
(3) all strings of integers $1, 2, \ldots, k$ beginning with 1.

The meet operation is defined by $x \sqcap y = \mathbf{0}$ whenever $x \neq y$. Thus, if $x \sqsubseteq y$, then either $x = y$ or $x = \mathbf{0}$.

The set F_{AB} of operations on L_{AB} includes

(1) the identity operation e on L_{AB};
(2) operations f_i, for $1 \leq i \leq k$, defined by as follows:
    (2a) if α is a string of integers beginning with a 1, then $f_i(\alpha) = \alpha i$,
    (2b) $f_i(\mathbf{0}) = (\mathbf{0})$, and
    (2c) $f_i(\$) = \$$;
(3) the operation g defined by as follows:
    (3a) for strings $\alpha = 1\, i_1 i_2 \ldots i_m$, $g(\alpha) = \mathbf{0}$ iff $1 i_1 i_2 \ldots i_m$ is a solution to instance AB of MPCP, and $g(\alpha) = \$$ otherwise;
    (3b) $g(\mathbf{0}) = \mathbf{0}$;
    (3c) $g(\$) = \$$;
(4) the operation h defined by $h(x) = 1$ (i.e., the string consisting of 1 alone) for all $x \in L_{AB}$;
(5) all operations constructed from the above by composition.

LEMMA 9.2 $D_{AB} = (L_{AB}, \sqcap, F_{AB})$ *as constructed above is a monotone framework.*

Proof It is easy to see that $(L_{AB}, \sqcap)$ is a semilattice of finite length, because of the definition of $\sqcap$.

To show that F_{AB} is a monotone operation space associated with L_{AB}, we shall only show monotonicity, as the other conditions are easily met.

By Lemma 9.1, it suffices to show that if $x \sqsubseteq y$ for $x, y \in L_{AB}$, then

(1) $f_i(x) \sqsubseteq f_i(y)$ for $1 \leq i \leq k$,
(2) $g(x) \sqsubseteq g(y)$, and
(3) $h(x) \sqsubseteq h(y)$.

[5] See Hopcroft and Ullman [1969], for example.

Since $h(x) = h(y) = 1$, (3) is immediate. For this meet operation, $x \sqsubseteq y$ implies either $x = y$ or $x = \mathbf{0}$. In the former case (1) and (2) are immediate. In the latter case, $f_i(x) = \mathbf{0}$ and $g(x) = \mathbf{0}$, so $f_i(x) \sqsubseteq f_i(y)$ and $g(x) = g(y)$ follow. □

THEOREM 9.4 *There does not exist an algorithm W with the following properties.*

(1) *The input to W is*
 (i) *algorithms to perform meet and application of operations on semilattice elements for some monotone framework and*
 (ii) *an instance I of the framework.*
(2) *The output of W is the MOP solution for I.*

Proof We show that if W exists, then MPCP is decidable. If W exists, then for each instance AB of MPCP we can apply W to the monotone framework D_{AB} constructed as above with instance I as in Figure 9.5. The set of all strings entering the top of the node associated with operation g is the set S of all strings of integers $1, 2, \ldots, k$ beginning with 1. If the instance AB or MPCP has no solutions, then given any $x \in S$, $g(x) = \$$. In that one

$$\bigsqcap_{x \in S} g(x) = \$.$$

If there exists a $y = 1 i_1 i_2 \ldots i_r$ such that $a_1 a_{i_1} \ldots a_{i_r} = b_1 b_{i_1} \ldots b_{i_r}$, then $g(y) = \mathbf{0}$. Hence $\bigsqcap_{x \in S} g(y) = \mathbf{0}$.

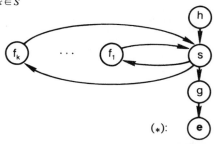

Figure 9.5 Instance *I* of *L* associated with instance *AB* of MPCP.

Thus, the MOP solution to instance I at point (*) is easily seen to be $\$$ iff the instance AB of the MPCP has no solution, and $\mathbf{0}$ otherwise. So if W exists, we could solve MPCP. □

We now show that there is a particular framework, namely the constant propagation framework CONST, for which no algorithm exists to compute, for all instances, the MOP solution.

The General Iterative Algorithm

For MPCP, there is no loss in generality in assuming that each a_i and b_i is in $\{1,2\}^+$ instead of $\{0,1\}^+$. In the following we show how an instance of MPCP can be transformed into a particular constant propagation problem.

DEFINITION Let x be any string in $\{1,2\}^+$. We define *the decimal value of x*, denoted $\text{decval}(x)$, to be the base 10 integer interpretation of x. For example, $\text{decval}(\text{'12'})$ is twelve (i.e., 12 base 10), and $\text{decval}(\text{'121'})$ is one hundred twenty-one (i.e., 121 base 10). Furthermore, if x any y are any strings in $\{1,2\}^+$, then $\text{decval}(xy) = \text{decval}(x) \cdot 10^{|y|} + \text{decval}(y)$, where $|y|$ denotes the length of string y. For example, if $x = \text{'12'}$ and $y = \text{'121'}$, then $xy = \text{'12121'}$, $|y| = 3$, and $\text{decval}(xy) = \text{decval}(\text{'12121'}) = 12 \times 10^3 + 121 = 12{,}121$ in base 10.

THEOREM 9.5 *There exists no algorithm to compute, for all instances of the constant propagation framework CONST, the MOP solution.*

Proof Let AB be any instance of MPCP, where $A = a_1, a_2, \ldots, a_k$ and $B = b_1, b_2, \ldots, b_k$, and $a_i, b_i \in \{1,2\}^+$. Then instance AB of MPCP has no solution iff

$$(C, 1) \in \bigsqcap_{P \in \text{PATH}(m)} f_P(\mathbf{0}),$$

where m is the designated node in the instance of CONST shown in Figure 9.6. We assume here that for integers x and y,

(i) $x \neq y$ iff $(x-y)/(x-y) = 1$, and
(ii) $x = y$ implies $(x-y)/(x-y) = 0/0 \neq 1$. □

9.5 The General Iterative Algorithm

DEFINITION Let $I = (G, M)$ be an instance of a monotone framework $D = (L, \sqcap, F)$. Let f_i denote $M(i)$, the operation associated with node i, and let the nodes of G be numbered from 1 to n by rPOSTORDER. The *maximum fixed point (MFP) solution* of I is defined as maximum fixed point of the following equations:

$$X[1] = (\mathbf{0}), \quad X[i] = \bigsqcap_{j \in \text{PRED}(i)} f_j(X[j]) \quad \text{for } 2 \leq i \leq n. \quad (9.1)$$

OBSERVATION 9.1 *The above equations have a solution that is a fixed point. Furthermore, it makes sense to talk about the maximum fixed point solution to the above equations.*

Proof Omitted. □

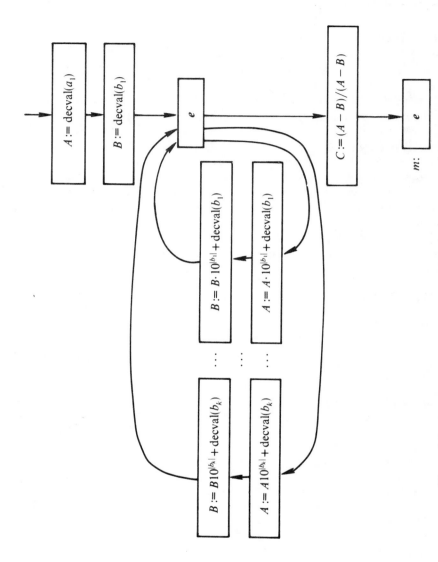

Figure 9.6 An instance of CONST associated with the instance AB of MPCP.

The General Iterative Algorithm

Kildall [1973] provided a general iterative algorithm for distributive frameworks. He proved that his algorithm converges to the MFP solution of a distributive framework, and, for distributive frameworks the MFP solution is equal to the MOP solution. Kam and Ullman [1975] proved that when Kildall's iterative algorithm is applied to a monotone framework, it converges to the MFP solution of that framework.

We now present variations of Kildall's general algorithm for monotone frameworks. Three of these are generalizations of the corresponding (worklist, round-robin, and node listing) iterative algorithms in Chapter 7. The fourth variation we present is due to Kam and Ullman [1975], and it does no worse than Kildall's algorithm and better on some problems. We believe that the worklist and round-robin versions of Kildall's general iterative algorithm (or the Kam-Ullman variant) will prove to be the most widely used algorithms for data flow analysis problems in practice, because they are so easy to program. The node listing version is messy because, in practice, precomputation of a suitable node listing may be involved.

ALGORITHM 9.1 *The general iterative algorithm (worklist version).*

INPUT: An instance $I = (G, M)$ of a monotone framework $D = (L, \sqcap, F)$, where $G = (N, A, s)$ is an n-node flow graph. Let $N = \{1, 2, \ldots, n\}$, and assume that $s = 1$.

We assume that if $(L, \sqcap)$ does not have a one element, an artifical one element can be added to L.

OUTPUT: An array $A[1:n]$ of semilattice elements.

METHOD: See procedure K1 in Figure 9.7. Array $A[1:n]$ holds semilattice elements and is global to K1. $A[i]$ is associated with the top of node i, $1 \leq i \leq n$. It is very important to note the use of $\not\sqsupseteq$ in procedure K1, and that $\sqsubseteq$ and $\not\sqsupseteq$ are *not* the same in a semilattice, in general. (Why?) □

THEOREM 9.6 *Given an instance $I = (G, M)$ of a monotone framework $D = (L, \sqcap, F)$, Algorithm 9.1 halts and the obtained array A is the MFP solution of I.*

Proof location Kam and Ullman [1975]. □

THEOREM 9.7 *Given an instance $I = (G, M)$ of a distributive framework $D = (L, \sqcap, F)$, at completion of Algorithm 9.1 the obtained array A is the MOP solution of I.*

Proof location Kildall [1973]. □

Theorems 9.6 and 9.7 also hold for Algorithms 9.2 and 9.3.
Here is a round-robin version of Kildall's general iterative algorithm.

procedure K1
  set W /* worklist */
  semilattice element B
  integer i, j
  /* Initialize array A. */
  $A[1] := \mathbf{0}$
  for $i := 2$ **to** n **by** 1 **do** $A[i] := \mathbf{1}$ **endfor**
  /* Initialize W. */
  /* Visit each node at least once (in case the **1** is artifical). */
  for $i := 1$ **to** n **by** 1 **do**
    $B := f_i(A[i])$
    $W := W \cup \{(j, B) | j \in \text{SUC}(i)\}$
  endfor
  /* Process worklist W. */
  while $W \neq \varnothing$ **do**
    Select and delete any pair (i, B) from W.
    if $B \not\sqsupseteq A[i]$ **then**
      $A[i] := A[i] \sqcap B$
      $B := f_i(A[i])$
      $W := W \cup \{(j, B) | j \in \text{SUC}(i)\}$
    endif
  endwhile
return

Figure 9.7   **The general iterative algorithm (worklist version).**

ALGORITHM 9.2   *The general iterative algorithm (round-robin version).*

INPUT: Same as Algorithm 9.1 except the nodes are numbered by rPOSTORDER.

OUTPUT: Same as Algorithm 9.1.

METHOD: See procedure K2 in Figure 9.8, where again the array $A[1:n]$ of semilattice elements is global to K2, and $A[i]$ is associated with the top of node i, $1 \leq i \leq n$. If the semilattice L contains a one element **1**, then the **for**-loop in the initialization step of procedure K2 can be changed to

'**for** $i := 2$ **to** n **by** 1 **do** $A[i] := \mathbf{1}$ **endfor** '.

Here is the node listing version of the general iterative algorithm. To simplify the algorithm we assume that L has a one element. (Otherwise, we can introduce an artificial **1**.)   □

ALGORITHM 9.3   *The general iterative algorithm (node listing version).*

INPUT AND OUTPUT: Same as Algorithm 9.2.

METHOD: Associate $A[i]$ with the top of node i, $1 \leq i \leq n$.

Initialization: $A[s] := \mathbf{1}$

The General Iterative Algorithm

```
procedure K2
  semilattice element TEMP
  integer i
  proposition CHANGE
  /* Initialize array A. */
  A[1] := 0
  for i := 2 to n by 1 do A[i] := ⊓_{g ∈ PRED*(i)} f_g(A[g]) endfor
  (PRED*(i) = {q | q ∈ PRED(i) and q < i in rPOSTORDER}.)
  /* Iterate. */
  CHANGE := true
  while CHANGE do
    CHANGE := false
    for i := 2 to n by 1 do
      TEMP := ⊓_{q ∈ PRED(i)} f_q(A[q])
      if TEMP ≠ A[i] then
        CHANGE := true
        A[i] := TEMP
      endif
    endfor
  endwhile
return
```

Figure 9.8 The general iterative algorithm (round-robin version).

Iteration step: $A[x] := \mathbf{0}$ for all $x \in N$, $x \neq s$. Visit nodes other than s in order $x_1, x_2, \ldots$ (not necessarily fixed in advance). We visit node x by setting

$$A[x] := \bigsqcap_{p \in \text{PRED}(x)} f_p(A[p]).$$

The sequence $x_1, x_2, \ldots$ has to satisfy the following condition: if there exists a node $x \in N - \{s\}$ such that

$$A[x] \neq \bigsqcap_{p \in \text{PRED}(x)} f_p(A[p])$$

after we have visited node x_u in the sequence, then there exists an integer $v > u$ such that $x_v = x$. Also, if

$$A[x] = \bigsqcap_{p \in \text{PRED}(x)} f_p(A[p]), \quad \text{for all } x \neq s,$$

then the sequence will eventually end. □

There is another variation of the iterative algorithm due to Kam and Ullman that gives better solutions for some monotone frameworks than do Algorithms 9.1, 9.2, and 9.3. Here is a node listing version of it for comparison with Algorithm 9.3.

ALGORITHM 9.4 *Improved version of Algorithm* 9.3.
INPUT AND OUTPUT: Same as Algorithm 9.3, except array H is output.
METHOD: Associate $B[i]$ with the bottom of node i, and $H[i]$ with the top, $1 \leq i \leq n$.
  Initialization: $B[s] := f_s(\mathbf{0})$
  $B[x] := \mathbf{1}$, for all $x \in N$, $x \neq s$.
  Iteration step: Visit nodes in "legal order" by setting

$$B[x] := \bigsqcap_{p \in \text{PRED}(x)} f_x(B[p]).$$

By "legal order" we mean the sequence $x_1, x_2, \ldots$ is as qualified in Algorithm 9.3.
  Final step: $H[s] := \mathbf{0}$
  $H[x] := \bigsqcap_{p \in \text{PRED}(x)} B[p]$, for all $x \in N$, $x \neq s$. □

Example 9.6 Consider the instance of CONST given in Figure 9.4. We invite the reader to verify that the Kam-Ullman variant (Algorithm 9.4) happens to produce the MOP solution for this instance, whereas Kildall's algorithm (Algorithm 9.1 for example) does not.

THEOREM 9.8 *Given an instance* $I = (G, M)$ *of a monotone framework* $D = (L, \sqcap, F)$ *as input to Algorithm* 9.4:

(1) *Algorithm 9.4 will eventually halt. The result array H is unique, independent of the order in which the nodes are visited, and*

$$H[i] \sqsubseteq \bigsqcap_{Q \in \text{PATH}(i)} f_Q(\mathbf{0}), \quad 1 \leq i \leq n.$$

(2) *The result array B is the MFP solution of the equations*

$$X[i] = f_1(\mathbf{0})$$

$$X[i] = \bigsqcap_{j \in \text{PRED}(i)} f_i(X[j]), \quad 2 \leq i \leq n. \tag{9.2}$$

A Dominator Algorithm for an Arbitrary Flow Graph

(3) *If the array A is the result of applying Algorithm 9.3 to $I = (G, M)$, then $A[i] \sqsubseteq H[i]$ for all i, $1 \leq i \leq n$.*

Proof location Kam and Ullman [1975]. □

Note the difference between Equations (9.1) and (9.2). In the former, $X[i]$ is associated with the top of node i, whereas in the former $X[i]$ is associated with the bottom of node i.

9.6 A Dominator Algorithm for an Arbitrary Flow Graph

In this section we show that the general iterative algorithm (e.g., round-robin version) can be used to compute the dominators of each node in a flow graph. We do this by simply designing a distributive framework in which the set of dominators of each node is the MOP solution. By Theorem 9.7 then, we have an algorithm to compute dominators for an arbitrary flow graph.

Let $G = (N, A, s)$ be a flow graph. Let $L = 2^N$, the set of all subsets of N, let $\sqcap$ denote set intersection, and let $\sqsubseteq$ be set containment. It is easy to verify that $(L, \sqcap)$ is a semilattice of finite length with both a zero element $\emptyset$ and a one element N.

Let F be defined as the set of all operations on L such that the identity operation e is in F, f_i is in F for each node i such that for each subset S of N we have $f_i(S) = \{i\} \cup S$, and F is closed under composition.

LEMMA 9.3 $(L, \sqcap, F)$ *as defined above is a distributive framework.*

Proof All that remains to be done is to verify conditions C4 and C5. We leave C4 to the reader and consider C5.

Clearly e is distributive, because $e(S_1 \cap S_2) = S_1 \cap S_2 = e(S_1) \cap e(S_2)$ by the definition of an identity operation. Also, the fact that f_i is distributive for any i follows from $f_i(S_1 \cap S_2) = \{i\} \cup (S_1 \cap S_2) = (\{i\} \cup S_1) \cap (\{i\} \cup S_2) = f_i(S_1) \cap f_i(S_2)$. We can now prove a result analogous to Lemma 9.1 except with distributive operations. Thus, each $f \in F$ is distributive, satisfying C5. □

LEMMA 9.4 *Consider an instance I of the distribute framework defined above in which f_i is assigned to node i, where $(\forall S \in L)[f_i(S) = \{i\} \cup S]$, $1 \leq i \leq n$. The MOP solution of i gives the set of dominators for each node i.*

Proof Suppose $j \in \text{DOM}(i)$; that is, j dominates i. Then j is on all paths from s to i. If Q is any path from s to i, then $j \in f_Q(\mathbf{0})$, because $f_Q(\mathbf{0})$ is just the set of nodes in path Q excluding i. By the definition of dominance and the meaning of $f_Q(\mathbf{0})$, j is in the intersection of $f_Q(\mathbf{0})$ over all paths Q from s to i. □

COROLLARY 9.1 *Algorithm 9.2, for example, can be used to compute dominators.*

Proof By Theorem 9.7. □

Exercises

9.1 Formulate range analysis of variables as a data flow analysis problem.

9.2 Given a programming language with set variables, formulate a data flow analysis problem that determines what operations are performed on such variables. Can this information be used for compile-time data structure selection?

Bibliographic Notes

Kildall [1972, 1973] first introduced the notion of a semilattice-theoretic model for data flow analysis. He observed that in a data flow analysis problem, the relevant information can be nicely modeled by a semilattice with a zero element and of finite length, and the effect of the information due to a basic block can be viewed as an operation on the semilattice. The MOP solution is the desired solution for a framework. Also, the constant propagation and "structured partition" frameworks he proposed for code improvement are quite different from existing bit vector frameworks.

Kildall presented a general iterative algorithm (Algorithm 9.1) for distributive frameworks, and he proved that it converges to the MFP solution of the framework, independent of the order in which nodes are visited. Furthermore, for a distributive framework the MFP solution is equal to the MOP solution.

Kam and Ullman [1975] introduced the abstraction of a monotone framework, and proved many key results about such frameworks including the following. The MFP solution exists for every instance of a monotone framework, and it can be obtained from Kildall's iterative algorithm. However, whenever the framework is monotone but not distributive, there are instances in which the (desired) MOP solution differs from the MFP. The MFP solution (to either Equation (9.1) or (9.2)) is always $\sqsubseteq$ the MOP solution. Finally, the MOP problem is undecidable. The definitions in this chapter, the proof of Theorem 9.4, and Algorithms 9.2, 9.3, and 9.4 come from Kam and Ullman [1975]. Theorem 9.5 comes from Kam [1975]. The dominator algorithm in Section 9.6 appears in Allen and Cocke [1972].

Graham and Wegman [1976] and Fong, Kam, and Ullman [1975] have considered fast frameworks. Kam and Ullman [1975] have found necessary and sufficient conditions for Algorithm 9.2 to converge within the bound of the bit vector frameworks (Lemma 7.2).

Constant propagation and structured partition frameworks are discussed in Kildall [1972, 1973]. A type determination framework is given in Tenenbaum [1974]. Schwartz [1975] contains a framework for gleaning information helpful for automatic data structure selection. Fong, Kam, and Ullman [1975] present a framework for induction variable anaysis.

Backhouse [1976] considers some elimination methods for the solution of the equations (9.1). Tarjan [1976] defines a hierarchy of data flow analysis problems that require increasingly longer node-listings. Fong [1977] considers a code improvement problem that does not fit the semilattice-theoretic mold. See Harrison [1975] for a solution to Exercise 9.1, and Schwartz [1975] for information about Exercise 9.2.

Part IV
A SIMPL Code Improver

Chapter 10

A MODEST QUAD IMPROVER FOR SIMPL-T

The primary application of flow analysis of computer programs is code improvement. Some ideas on the design and implementation of a modest code improver for the structured programming language SIMPL-T are presented in this chapter.

There are two major purposes for this chapter. First, a method for flow analysis different from that developed in Chapters 1 to 9 is presented: top-down recursive descent flow analysis. The scenario heretofore presented is to construct a call graph for the program and a flow graph for each procedure; then decide on one or more strategies for interprocedural analysis; and then select an intraprocedural data flow analysis algorithm (e.g., the iterative algorithm, Ullman's algorithm, or the Graham-Wegman algorithm) to collect information that is necessary for code improvement. One important tacit assumption of this scenario is that the control flow (i.e., loop structure) of a procedure is arbitrary rather than highly restricted, and thus intraprocedural data flow analysis algorithms that work for arbitrary control flow are necessary. For a structured programming language such as SIMPL-T, the control flow (and specifically the loop structure) is highly restricted. Consequently, a less general intraprocedural data flow analysis algorithm suffices in which the construction of flow graphs is obviated. (By 'less general' here we mean that the algorithm does not work for programs with arbitrary loop structure.) The second purpose of this chapter is to illustrate, by an extended example, some typical interprocedural analysis considerations and the interplay between these and the intraprocedural analysis selected.

The purpose of this chapter is not to present a case study that reviews and selects from among the techniques in Chapters 1 to 9 (despite the fact that such a case study would be very valuable). Rather, it is to present *another* method for flow analysis. The techniques presented in earlier chapters are necessary for programming languages such as FORTRAN and PL/I that have **go to** statements and few control flow restrictions. However, we can tailor (in this case greatly simplify) the techniques selected to the idiosyncrasies of the programming language.

The material in this chapter is important to the practitioner because it demonstrates, by a specific example, how some intermediate-level code improvement can be performed on a structured programming language, in contrast to a language such as FORTRAN. Furthermore, an attempt is made to convince the reader that a modest code improver for a language such as SIMPL-T is not as difficult to construct as he may think.

First, motivation is given for intermediate-level code improvement, and the design goals for the improver herein described are explained. Next, relevant information about SIMPL-T is presented. An overview of the improver is then given. A description of the interprocedural analysis follows, with attention to the dynamic aliasing problem. Finally, the intraprocedural analysis used is described.

The material in this chapter comes from a study by J. B. Shaffer and the author.

10.1 Introduction

10.1.1 "Levels"

Program and code improvement transformations for a general, high-level programming language can be performed on any program representation at compile-time. Typically, a high-level source program is translated into an intermediate-level program representation such as a parse tree, two-address code (triples), or three-address code (quadruples), which is subsequently translated into a low-level program representation such as machine language.

Program improvement by source-to-source transformation is performed directly on the source program or its parse tree. There are two advantages of improvement at this level: first, the result of such transformations can be easily communicated back to the programmer in a language that he can understand; and second, high-level transformations such as factoring conditional statements out of loops are usually much easier at this level than at a lower level.

Code improvement on an intermediate-level program representation (such as quadruples) has some distinct advantages also. Standard language and machine independent transformations such as elimination of common subexpressions, elimination of redundant code, and removal of invariant code from loops are more thorough at this level than at the source level. As evidence, common code resulting from similar array references is usually impossible to eliminate at the source level. Also, because these improvements are relatively language and machine independent, there is no need to burden a code generator with this overhead and unnecessarily complicate the rewriting of the code generator when transporting the compiler to another machine. A second advantage of applying transformations to an

Introduction

intermediate-level program representation is that these transformations do not hinder the readability of the source program. As examples, addition of variables to hold common subexpressions at the source level, and in-line expansion of procedures (perhaps called only once) can quickly complicate program readability.

Finally, low-level code improvement, in addition to both high-level and intermediate-level improvement, is necessary because it is machine dependent, requires individual tailoring for different machines, and can't be done at a higher level. Large gains can be made in scrutinizing low-level code, despite the fact that this problem generally requires *ad hoc* analysis and has consequently thwarted elegant mathematical models.

Clearly, program or code improvement transformations are useful at many phases of the compilation process. In fact, it is now generally acknowledged that, for a thorough job, program and code improvement transformations should be attempted at all three (high, intermediate, and low) levels of program representation. Most research, however, has been directed at the intermediate level because there has been some success in mathematically modeling problems at this level (with flow graphs). Many of the contributions are of a specific nature, and little work has been done in synthesizing ideas into an integrated package. The task of writing a code improver, even for the well studied intermediate level, is portrayed as a gargantuan one. We believe that it is not at all as difficult as currently believed to write such a code improver, provided modest goals are set and a well-designed language is the target for improvement. (The task may in fact be gargantuan for a programming language such as PL/I.)

10.1.2 Design Goals

Our thesis is, that given a structured programming language whose intermediate representation is relatively high-level quadruples (i.e., the source-level control flow structures remain easily identifiable in the quadruples), a modest but effective "quad improver" employing interprocedural analysis can be easily implemented with a slightly modified recursive descent design. Thus, we are restricting our attention to intermediate-level code improvement on a simple, structured programming language. To prove this thesis, we have constructed a prototype implementation for SIMPL-T. In this chapter we outline the basic design of this improver.

The adjectives 'modest' and 'effective' in the previous paragraph need clarification. The word 'modest' indicates a limited and somewhat partial implementation (i.e., limited in size and power). Our design includes redundant expression elimination, removal of invariant loop computations, and in-line expansion of selected procedure calls. It does not include all known code improvement transformations. For example, code motion such as hoisting or sinking (factoring) common code from **IF-THEN-ELSE**[1] and **CASE** statements is not implemented, among others. In fact, it is not at all obvious how to integrate this ability into our design without significant additional data structure overhead. The word 'effective' reflects our goal of 10–20% average reduction in expected execution time and in the number of quadruples of large SIMPL-T programs written with the knowledge of what our improver can do.

Our three major goals were (1) good performance, (2) design simplicity, and (3) incorporation of some interprocedural analysis.

10.2 SIMPL-T

10.2.1 Comments about the SIMPL-T Compiler

SIMPL-T is a member of the SIMPL family of structured programming languages under development at the University of Maryland.

Two of the major design criteria for each compiler of a language in the SIMPL family are transportability and (of course) efficient code generation.

To achieve language transportability, (a) each compiler is written in its own language or an ancestor language; and (b) each compiler in the family translates source text into an intermediate text that is a file of high-level and relatively machine-independent quadruples (quads), which is input to an interchangeable code generator. The code generator is individually tailored to the host machine's instruction set and architecture.

To achieve efficient code generation, (a) the core language and each of its extensions have a very simple design; (b) the code generator, because of its individually tailored nature, can exploit instruction repertoire idiosyncrasies; and (c) a transportable and optional quad improver module just prior to code generation, as shown in Figure 10.1, is accommodated (and envisioned) by the original compiler design.

10.2.2 A Synopsis of SIMPL-T Language Features

SIMPL-T is a general purpose, structured programming language that is procedure-oriented and non-block-structured. A SIMPL-T program consists of a (possibly empty) sequence of global declarations, followed by a sequence of one or more separate (unnested though possibly **REC**ursive)

---

[1] We use capitalized boldface for SIMPL-T keywords. In contrast, recall that we use lower case boldface for Pidgin SIMPL keywords.

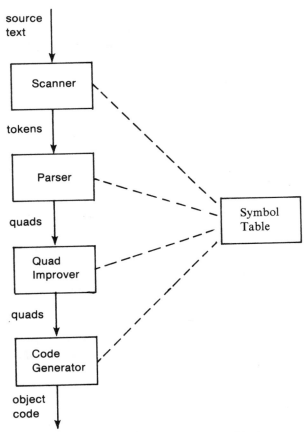

Figure 10.1 The SIMPL-T Compiler.

PROCedures and **FUNC**tions, followed by the keyword **START** then a procedure name. That procedure is called the *start procedure*, and execution begins with its first statement.

There are three scalar data types (scalars): **INT**eger, **STRING**, and **CHAR**acter. (SIMPL-R also has the scalar data type **REAL**.) Since there is no scalar data type 'logical', nonzero denotes 'true', and zero denotes 'false'. There is an extensive set of operations and intrinsic functions for each data type. For example, operations for the scalar data type **INT** include arithmetic, relational, logical, shift, bit, and partword. There are also type conversion intrinsic functions.

There are two data structures: **ARRAY** and **FILE**. Arrays are one-dimensional, consist of scalars of declared homogeneous type, and begin with element 0. Files (corresponding to sequential storage files) are sequences of (possibly heterogeneous) scalars.

Global declarations of variables (global variables, globals) are accessible from all procedures and can be initialized. Procedures may have local declarations of variables (local variables, locals) that are only accessible within the body of the procedure in which they are declared. Local declarations override global declarations of the same name, so that a global is not accessible within a procedure that redeclares it. Procedures may have parameters.

At run-time, local environments are destroyed and recreated between calls (rather than retained). The nonlocal environment of a procedure is determined by an implicit static specification; namely, the global environment.

Any procedure of a program may call any other procedure of that program. Only scalar and structure data types may be passed as actual arguments (actuals): not procedure names. Any formal parameter (formal) of a procedure is assumed to be call-by-value unless it is preceded by the keyword **REF** indicating call-by-reference; however, all structures are assumed to be call-by-reference. Procedures, therefore, may access globals, their locals, and their formals.

Statements in SIMPL-T include assignment (denoted by ':='), **IF, CASE, WHILE, CALL, EXIT, RETURN, ABORT, READ,** and **WRITE**. There are compound statements, but no block-structure. **IF, CASE,** and **WHILE** statements are terminated by the keyword **END**. (We use **ENDIF, ENDCASE,** and **ENDWHILE** instead.) There is no statement separator such as a semicolon. Comments (possibly nested) are delimited by /* and */.

Separately compiled programs may communicate via **EXT**ernal references and **ENTRY** points. Procedures and globals, preceded by the keyword **ENTRY**, are accessible to separately compiled programs. In order to access an **ENTRY** point of a separately compiled program, the identifier to be accessed must be declared and preceded by the keyword **EXT**. External procedure declarations must include a specification of the type and parameter-passing-mechanism of each formal. Only one of a group of separately compiled programs may specify a start procedure, which itself may be external. The remaining programs are called "nonexecutable programs", and are so designated by omitting the start procedure.

Example 10.1 Figure 10.2 shows two programs, one executable and the other nonexecutable to illustrate the use of **EXT** and **ENTRY**. (In SIMPL-T, the **WRITE** statement fills an output buffer that **SKIP** subsequently empties.)

```
INT A, B    /* globals */
EXT PROC SWAP (REF INT, REF INT)
PROC EXAMPLE           ENTRY PROC SWAP(REF INT X, REF INT Y)
   READ(A, B)              INT T   /* Local */
   WRITE(A, B, SKIP)       T := X
   CALL SWAP(A, B)         X := Y
   WRITE(A, B, SKIP)       Y := T
   RETURN                  RETURN
START EXAMPLE          START
        (i)                              (ii)
```

Figure 10.2 SIMPL-T **programs illustrating EXT and ENTRY features.**

There are two implementation features of SIMPL-T that affect quad improvement. First, if the value of a (binary) logical operation, such as **OR** or **AND**, can be determined from the first operand only, the second operand is not evaluated. Second, functions are assumed to have no side effects;[2] however, this is not enforced. The reason for the no-side-effects rule is due to the first feature above. If a function call involved in an unevaluated operand of a logical operation has a side effect, then that effect does not occur. In addition, it was presumed (by the language designers) that successive function calls whose actuals are unchanged are more easily detected and removed under the no-side-effects rule, although this is not the case in our design.

10.2.3 Quadruples of SIMPL-T

The SIMPL-T compiler translates the source text of a SIMPL-T program into an intermediate (internal) text, which is a file of relatively high-level quadruples. Each quadruple consists of (up to) four fields called ID, A, B, R.

In general, the ID field identifies the "operation", name, or type of the quad; the A and B fields are the two "operands" of the quad; and, the R field is the "result" field of the quad. Some quads have no A, B, and R fields (e.g., a unary operation quad has no B field). The A, B, and R words contain either a pointer to the symbol table, a "temporary number" or a number. Flags (bits denoting certain information) are used to distinguish among these.

Temporary variables, which we denote by $t1, t2, \ldots$, are used to hold intermediate computations.

---

[2] A *side effect* is a change in the value of a nonlocal variable. Thus in SIMPL-T a side effect is a change in the value of a global.

Example 10.2 The quads for $X := Y + Z$ are

$$(+, Y, Z, t1)$$

$$(:=, t1,, X)$$

where the meaning of each quad is self-evident.

When the parser produces quads, it numbers temporaries on a statement-by-statement basis. Whenever the contents of a temporary is to be saved for future use, the "quad save flag" (QSFLG) of the field containing that temporary is set. Thus, whenever a temporary is referenced such as in an A or B field and the QSFLG is not set, that temporary is reusable in the next statement.

Sometimes a temporary holds an address. When a field contains a temporary that must be referenced indirectly, the "quad indirect flag" is set. We denote an indirect reference of a temporary by preceding the temporary by a star.

Example 10.3 The quads for $A(I) := X + Y$ are

$$(ARRAYLOC, A, I, t1)$$

$$(+, X, Y, t2)$$

$$(:=, t2,, *t1)$$

where the first quad assigns the address of $A(I)$ to t1, the second quad computes $X + Y$, and the third quad assigns the value.

In addition to quads for assignment, array indexing, and (arithmetic, relational, logical, shifting, and bit vector) operations, there are quads for calling procedures, specifying actuals, and delimiting procedures and statements.

Each procedure is delimited by PROC and ENDPROC quads. There is also a RETURN quad. In procedures, the statement **RETURN** causes an immediate return to the caller. For functions, the A field of a RETURN quad identifies a variable or temporary that is the expression to be returned.

Each procedure call is delimited by CALL and ENDPARMS quads. Each actual parameter is represented by a PARM quad. The R field of a PARM quad is 1 for call-by-reference, and 0 for call-by-value. The R field of a CALL quad of a function identifies the temporary into which the function result is to be placed.

Examples of the quads corresponding to some statements are given below. We use curly braces to denote optional.

Example 10.4 The general form of an **IF** statement is given in Figure 10.3. Note that the **ELSE**-clause is optional. At run-time, the expression is evaluated, and the **THEN**-clause or **ELSE**-clause is executed upon a nonzero or zero result respectively.

IF ⟨expr⟩ THEN ⟨stmt list 1⟩ { ELSE ⟨stmt list 2⟩ } ENDIF	quads for ⟨expr⟩ (IF, ⟨expr result⟩,,) quads for ⟨stmt list 1⟩ { (ELSEIF,,,) quads for ⟨stmt list 2⟩ } (ENDIF,,,)

Figure 10.3 Quads for an IF statement.

Example 10.5 The general form of a **WHILE** loop is given in Figure 10.4 with the corresponding quads. The expression is evaluated prior to each potential execution of the loop, and a result of zero causes control to pass to the next statement following the loop. (Thus, the statement list may be executed zero times.) It is also possible to leave a **WHILE** loop by means of an **EXIT** statement, which transfers control to the statement immediately following that loop. Multiply nested loops may be **EXIT**ed by prefixing loops with a designator (identifier delimited by back slashes) and specifying the identifier in the **EXIT** statement (e.g., EXIT(LOOP2)). The quad for **EXIT** is (EXIT, ⟨designator⟩,,).

\⟨designator⟩\WHILE ⟨expr⟩ DO ⟨stmt list⟩ ENDWHILE	(WHILE, {⟨designator⟩},,) quads for ⟨expr⟩ (WHILETEST, ⟨expr result⟩,,) quads for ⟨stmt list⟩ (ENDWHILE,,,)

Figure 10.4 Quads for a WHILE loop.

Example 10.6 A hybrid mixture of a general and example **CASE** statement is shown in Figure 10.5. The expression is evaluated, and the statement list with corresponding descriptor corresponding to the value of the result is executed. If the result is not listed, then the **ELSE**-clause is executed if present. A similar form of **CASE** statement can be written with a character (but not a string) expression.

CASE ⟨expr⟩ OF	(INITCASE,,,)
\1\\2\	quads for ⟨expr⟩
⟨stmt list 1⟩	(CASETEST, ⟨expr result⟩,,)
\4\	(CASEID, 1,,)
⟨stmt list 2⟩	(CASEID, 2,,)
ELSE	quads for ⟨stmt list 1⟩
⟨stmt list 3⟩	(CASEEND,,,)
ENDCASE	(CASEID, 4,,)
	quads for ⟨stmt list 2⟩
	(CASEEND,,,)
	(CASEELSE,,,)
	quads for ⟨stmt list 3⟩
	(CASEEND,,,)
	(ENDCASE,,,)

Figure 10.5 An Example of the quads for CASE statement.

For some mysterious and unfortunate reason, SIMPL-T currently has no **go to** quad. The need for such a quad will be explained in Section 10.4.2.

10.3 Overview of the Quad Improver

We wish to exploit the simple control structure of languages such as SIMPL-T. In the case of SIMPL-T, it is possible to do so through direct scanning of the quad file. The high-level nature of the quads makes this possible, since the control flow structures are still identifiable and have not been replaced by indistinguishable jumps. Thus, we can perform a top-down recursive descent analysis, identifying and processing each control structure as it is encountered. As might be expected, this leads to a much less complicated algorithm than would be necessary for a primitive low-level program representation. For example, there is no problem in identifying "basic blocks", or in determining redundant expressions at a given point.

We now present an algorithm that performs redundant expression elimination, removes invariant code from loops, and expands selected procedures in-line for a structured programming language. Our improver has four phases. (See Figure 10.6.) The first phase (PASS1) scans the quad file, and collects data and control flow information needed for code improvement. The second phase (TC) performs transitive closures on some of the information gathered in the first phase. In the third phase (IN-LINE EXPANSION), we expand selected procedures in-line. Finally, in the fourth phase (PASS2) we process each procedure removing redundant expressions and invariant code from loops. Essentially, the first three

Interprocedural Analysis

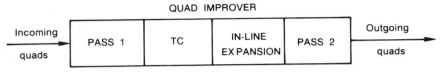

Figure 10.6   Phases of the quad improver.

phases perform interprocedural analysis, and the fourth phase performs intraprocedural analysis.

10.4   Interprocedural Analysis

In this section we elaborate on PASS1, TRANSITIVE CLOSURE, and IN-LINE EXPANSION.

Let us assume that our task is just to remove redundant expressions, assignments, function calls, and array indexing computations. One way to accomplish this is to process statements one by one, using and accumulating bookkeeping information such as (a) what expressions, function values, and addresses have already been computed and are currently available, and (b) what variables hold equivalent values. An appropriate representation for such information is an equivalence relation. Two basic operations on this data structure are query (when testing for redundancy), and update (to delete items no longer available or equivalent, and to insert items that have just beome available or equivalent).

Let us further assume that, because of its apparent simplicity, we shall use a (perhaps slightly modified) recursive descent approach to control the scanning of SIMPL-T quads.

The key problem of interprocedural analysis is how to handle **CALL** statements. There are two aspects of this problem:

(1) What is the effect of a **CALL** statement on the calling procedure? That is, how shall we update our equivalence relation upon hitting a **CALL** statement?
(2) What is the effect of a **CALL** statement on the called procedure? That is, if each procedure is to be analyzed once, what information is safe to assume for multiple calls?

To discuss these problems we need the concept of a call graph.

10.4.1   The Call Graph

SIMPL-T programs consist of a collection of procedures. An *internal procedure* is one that is present and available for analysis, whereas an *external procedure* is absent, and not available for analysis.

One representation of the calling relationships among the procedures of a program is a directed graph named a *call graph*. Recall (from Section 1.4.2) that there is a node in the call graph for each procedure, and a single arc from node A to node B iff A calls B one or more times. The call graph of a SIMPL-T program is easy to construct.

If the SIMPL-T program has a start procedure, then this procedure is called the *root* of the call graph. Nonexecutable SIMPL-T programs have no start procedure. We can depict the root of a call graph (if present) and each **ENTRY** procedure (if any are present) by a tailless arc entering the corresponding nodes. The call graph of a SIMPL-T program is not necessarily connected due to **ENTRY** procedures not called by any internal procedures.

Call graphs may contain cycles, due either to recursive procedures or to calling sequences in which the same procedures are called in a different order.

When processing a SIMPL-T program, we are completely uninformed as to the calling behavior of **EXT** procedures, and so we must make worst-case assumptions: each **EXT** procedure calls all other **EXT** and **ENTRY** procedures. We augment the call graph accordingly.

10.4.2 The Effect of a CALL Statement on the Calling Procedure

There are several ways to handle a **CALL** statement when processing the calling procedure. (For example, see Sections 1.5.2 and 1.6.) We consider two ways here.

One simple method is to expand the called procedure in-line, then continue processing as usual. This idea can sometimes be fruitful. However, it is clearly unsuitable for **EXT** procedures, because their bodies are unavailable; most **REC** procedures, because the expansion process if not monitored may never terminate; large procedures that have multiple calls, because of the unreasonable increase in program size; and **ENTRY** procedures, because **EXT** procedures might not be able to call them.

We have decided to expand procedures in-line that meet all of the following criteria:

(a) The procedure is called only once. (This simplifies the implementation somewhat.)
(b) The procedure is not the start procedure.
(c) The procedure is neither **EXT** nor **ENTRY**.
(d) The procedure has no **RETURN** statements embedded within it.

Interprocedural Analysis

The necessary information for (a) and (b) is easily gathered in PASS1 and stored in the nodes of the call graph. The information for (c) is stored in the symbol table.

It is not unreasonable to expand certain **REC** procedures in-line. For example, suppose A calls B calls C calls D calls B is the call graph, the only caller of C is B, and the only caller of D is C. Then it is okay to expand D into C and C into B.

To implement IN-LINE EXPANSION, we use the copy rule. Our algorithm first scans the quad file (QUADS1) and copies into a temporary file (TEMPFILE) all procedures that should be expanded according to our criteria. It then makes a second pass of QUADS1, copying quads into a new quad file (QUADS2). Calls of expandable procedures are replaced by the procedure body copied from TEMPFILE, and the original body of an expandable procedure is not copied from QUADS1 to QUADS2. During expansion, temporaries must be renumbered so as not to conflict with those in the surrounding code.

The process is a little more complicated than this, since an expandable procedure may call another expandable procedure. Thus we can be copying from TEMPFILE, find a CALL, and have to stop what we are doing to copy the new procedure before finishing the old one. A recursive set of procedures, keeping track of where we are and where we have been, are used for this.

It is important to note that if a procedure has a **RETURN** statement embedded within it, then the resulting expanded code requires an arbitrary forward **go to**.

It is our experience that at least 20% (i.e., one out of five) of the procedures of a large SIMPL-T program can be expanded in-line. Furthermore, the subsequent reduction in the number of quads is about 1–2%. We anticipate that execution timing experiments will prove in-line expansion to be an important component of a code improver.

A second method for handling a **CALL** statement is to make a pre-pass over the program text to determine the variables whose values may be modified (killed). With this "summary" information, the equivalence relation can be updated.

We assume that if a procedure may modify the value of a variable it will, and that if it may call another procedure it will. To implement these assumptions, PASS1 constructs an initial PKILL bit vector for each procedure, which has a bit position for each program variable. A bit is set to 1 iff the value of the corresponding variable may be modified in the given procedure. In TC we perform the appropriate transitive closure of the PKILL bit vectors using the call graph. We assume that each **EXT** procedure may modify the value of each **EXT** and **ENTRY** variable.

10.4.3 Summary of the Pre-pass

PASS1, the pre-pass, not only constructs the call graph and initial PKILL bit vectors, but also gathers some other information.

It is also useful to know those variables whose values may be modified by **WHILE** loops, as we shall see in PASS2. We handle **WHILE** loops by using bit vectors too. Each loop is numbered and has a bit vector WKILL, similar to PKILL for procedures. There is also a set of bit vectors PCIW, which record procedure calls within **WHILE** loops. After the transitive closure of the PKILL bit vectors has been performed in TC, we can then complete the WKILL of each loop by ORing in the PKILL of those procedures that it calls. Nested loops are handled as follows: The WKILL and PCIW bit vectors of the current innermost loop are updated as we pass through the quads. Whenever we reach the end of an inner loop, we OR its two bit vectors to those of the surrounding one. Thus each bit vector is completed in one top-down pass.

We also number **IF** statements, corresponding to bits in the bit vector HASANELSE. A bit is set to 1 iff the corresponding statement has an **ELSE** clause.

The program outline in Figure 10.7 summarizes the functions of PASS1.

10.4.4 Transitive Closure

In TC, transitive closures are performed on the PKILL and WKILL bit vectors, and also on the aliasing information (to be discussed in the next section). We use the same transitive closure algorithm for both the PKILL bit vectors and the aliasing information. However, since the initialization of aliasing information is quite interesting, we postpone the discussion of its transitive closure until the next section.

The first transitive closure is performed on the PKILL bit vectors, with the help of the call graph. From the transitive closure of the PKILLs we then compute the transitive closure of the WKILLs. Transitive closure implied by this type of graph can be easily implemented through a form of depth-first search that looks for strongly connected components (SCCs). If each SCC is replaced by a single node, then the resulting graph is acyclic. In the call graph, procedures in an SCC may all call each other (perhaps indirectly) and hence the PKILL of any procedures is the bit vector union of all those in the SCC. These facts allow us to use depth-first search to:

(a) Find SCCs in depth-first order and form the bit vector union of the PKILLs of their members.
(b) Form the bit vector union of the PKILL of a called SCC to a calling SCC as we retreat back up the graph.

Interprocedural Analysis

```
rec proc PASS1
while more quads in file do
  read a quad
  case quad of
    \assignment\
      Set corresponding bit in current PKILL.
      Set corresponding bit in current WKILL
        if we are within a while loop.
    \WHILE\
      Number loop (and start using new WKILL, PCIW).
      call PASS1 recursively.
    \ENDWHILE\
      OR bit vectors of current loop to outer loop, if one exists.
      return from recursive call.
    \IF\
      Number statement.
      Stack number for nested IF processing.
    \ELSEIF\
      Set corresponding bit in HASANELSE.
    \ENDIF\
      Unstack IF number.
    \CALL\
      Add new node and/or arc to call graph if
        not already there. Set corresponding PCIW
        bit if we are inside a loop.
      Recursively process parameters; add reference
        parameters to current
        WKILL if it exists.
    \new procedure\
      Add node to call graph if necessary.
      Begin using a new PKILL bit vector.
  endcase
endwhile
```

Figure 10.7 Summary of PASS1.

This can be done in $O(\text{arcs})$ bit vector unions. Once the PKILL bit vectors are completed, they are ORed to WKILL bit vectors whenever a procedure call occurs within a **WHILE** loop (as determined from the PCIW bit vectors).

ALGORITHM 10.1 *Transitive closure of* PKILLs *via depth-first search.*

Initially, mark all nodes 'unvisited'.
Stack S is empty.
COUNT is 0.

Arrays DFNUMBER, LOWLINK, and SCCROOT are global.
while ∃ an unvisited node X **do**
  call DFS(X)
endwhile

The DFS routine is given in Figure 10.8. □

---

rec proc DFS(**int** X)
```
*------------------------*
*    Perform depth-first search.       *
*    Find strongly connected components. *
*    Form appropriate set unions.      *
*------------------------*
```
  int Y, Z
  Mark X 'visited'.
  COUNT := COUNT + 1
  DFNUMBER(X) := COUNT
  LOWLINK(X) := COUNT
  Push X onto stack S.
  for each successor Y of X **do**
    if Y is not yet visitied **then**
      call DFS(Y)
      LOWLINK(X) := **min**(LOWLINK(X),LOWLINK(Y))
    else
      if (DFNUMBER(Y) < DFNUMBER(X))
        and not SCCROOT(Y) then
        /* back arc or cross arc */
        LOWLINK(X) := **min**(DFNUMBER(Y), LOWLINK(X))
      endif
    endif
    if LOWLINK(Y) = DFNUMBER(Y) **then**
    /* Y is the root of an SCC */
      PKILL(X) := PKILL(X) **bitvectorOR** PKILL(Y)
    endif
  endfor
  if LOWLINK(X) = DFNUMBER(X) **then**
  /* X is the root of an SCC */
    repeat
      Pop stack S to obtain Z.
      if $X \ne Z$ **then**
        PKILL(X) := PKILL(X) **bitvectorOR** PKILL(Z)
      endif
      SCCROOT(Z) := X
    until $Z = X$
  endif
return

Figure 10.8 Depth-first search routine for transitive closure of PKILL bit vectors.

10.4.5 Aliasing

We now consider the effect of a **CALL** statement on the called procedure. The fundamental problem here is called dynamic *aliasing*, and is due to syntactically distinct names representing the same or overlapping storage areas at run-time. Aliasing may cause both side effects and surreptitious modification of the values of local variables.

There are two aspects of aliasing that concern us: how to detect it, and how to cope with it.

We say that variables A and B are *aliases* iff, whenever there is an assignment of a value to A, that value is also assigned to B, and vice versa.

In SIMPL-T, aliasing may only be introduced at run-time by procedure calls (if at all) due to the call-by-reference parameter passing mechanism. Since aliasing in SIMPL-T is dynamic, different calls to a procedure may induce distinct aliasing relationships. Thus, aliasing relationships are relative to a particular call or sequence of calls. There are three specific situations in which aliasing is caused in SIMPL-T. We list them as observations.

OBSERVATION 10.1 *If a global G is passed as a call-by-reference parameter X to procedure P, then, relative to that call, G and X are aliases in P.*

OBSERVATION 10.2 *If a call-by-reference parameter X in procedure P is passed as a call-by-reference parameter Y to procedure Q, then relative to the particular sequence of calls that currently has reached Q, each global G, which is an alias of X in P, is now an alias of Y in Q.*

OBSERVATION 10.3 *If a procedure P has two or more call-by-reference parameters, and if in some call of that procedure two or more of them are passed the same actual parameter A, then they are aliases of each other (relative to that call) in P.*

Here are three examples that illustrate aliasing.

Example 10.7 Consider Figure 10.9, where procedure P has **REF** formal F. There are two calls of P in MAIN. Global G and formal F are aliases in P relative to the first call of P in MAIN, and global H and formal F are aliases in P relative to the second call of P in MAIN.

Example 10.8 Suppose G is a global, procedure P has local L, procedure Q has two **REF** formals X and Y, and P calls Q with identical actuals as in Figure 10.10. Then X and Y are aliases in Q relative to the first call of Q in P. X, Y, and G are aliases in Q relative to the second call of Q in P.

```
INT G = 1,  /* global */
    H = 5   /* global */
PROC MAIN
  CALL P(G)
  CALL P(H)
  RETURN
PROC P(REF INT F)
  WRITE(F, G, H, SKIP)
  F := F + 1
  WRITE(F, G, H, SKIP)
  G := G + 1
  WRITE(F, G, H, SKIP)
  H := H + 1
  WRITE(F, G, H, SKIP)
  RETURN
START MAIN
```

Figure 10.9

```
INT G = 1   /* global */
PROC MAIN
  CALL P
  RETURN
PROC P
  INT L   /* local */
  L := 6
  CALL Q(L, L)
  CALL Q(G, G)
  RETURN
PROC Q(REF INT X, REF INT Y)
  WRITE(X, Y, G, SKIP)
  X := X + 1
  WRITE(X, Y, G, SKIP)
  Y := Y + 1
  WRITE(X, Y, G, SKIP)
  G := G + 1
  WRITE(X, Y, SKIP)
  RETURN
START MAIN
```

Figure 10.10

Interprocedural Analysis

Example 10.9 Suppose procedure P has one **REF** formal X, procedure Q has one **REF** formal Y, and P calls Q with actual X as in Figure 10.11. Then any alias of X in P that is a global is an alias of Y in Q relative to the MAIN calls P calls Q sequence. Since Q and X are aliases in P, then G and Y are aliases in Q.

```
INT G = 1
PROC MAIN
   CALL P(G)
   RETURN
PROC P (REF INT X)
   WRITE(G, X, SKIP)
   CALL Q(X)
   WRITE(G, X, SKIP)
   RETURN
PROC Q (REF INT Y)
   WRITE(G, Y, SKIP)
   G := G + 1
   WRITE(G, Y, SKIP)
   Y := Y + 1
   WRITE(G, Y, SKIP)
   RETURN
START MAIN
```

Figure 10.11

Now we consider the problem of how to cope with aliasing.

Let S denote the set of scalar globals and scalar **REF** formals. In SIMPL-T, we are only concerned with aliasing among variables in S. The scalar aliasing produced by a procedure call is named an *aliasing pattern*. An aliasing pattern is simply an equivalence relation among names, telling which names denote the same storage.

Since distinct calls of a procedure P at run-time may produce distinct aliasing patterns, what assumptions should be made when analyzing P at compile-time? Or, should P be analyzed more than once at compile-time?

Several methods have been proposed. One method is to process each procedure once assuming that all elements of S are aliases. [Actually, the scope of aliasing can be limited by block structure (if present) and type compatibility (if strong typing is present).] Another method is to first determine all aliasing patterns for each procedure. Then, process each procedure once assuming as aliasing for each procedure an "aggregate" of the aliasing patterns for that procedure. A third method is to analyze each procedure (symbolically) as many times as there are distinct aliasing patterns for it.

We chose to implement the second method above. Thus, after some preprocessing (i.e., PASS1, TC, and IN-LINE EXPANSION), we process each procedure once and use aggregate aliasing patterns when analyzing a procedure. The decision to limit the analysis of each procedure here to once is an attempt to limit improver time.

We now describe an algorithm to form aggregate aliasing patterns. The idea is to record the aliasing pattern of each **CALL** statement, implicitly representing transitive effects, and then to replace implicit transitive effects by an explicit transitive closure. Let us assume that aggregate aliasing patterns are represented by lists.

An alias list is established for each variable in S. Initially, we fill alias lists with two types of elements from S: "atoms", and "pointers to other

---

proc INITIALIZE
/* Initialize alias lists of **REF** formals. */
  for each **CALL** statement **do**
    if the situation in Ovservation 10.1 occurs **then**
      Place G on X's list.
      * That is, G and X are alias within
      X's procedure. *\
    endif
    if the situation in Observation 10.2 occurs then
      Place X on Y's list.
      * That is, every global that is an alias of X
      should be an alias of Y in Y's procedure. *\
    endif
    if the situation in Observation 10.3 occurs **then**
      for each pair X, Y of **REF** formals that
      have identical actuals **do**
        Place *Y on X's list, and place *X on Y's list.
      endfor
      /* When we put *Y on X's list, for example.
        this means that Y is an alias of X, and
        every global that is an alias of Y
        should also be an alias of X. */
      if the identical actual is a global G **then**
        Place G on both X's and Y's list.
      endif
    endif
  endfor
return

      Figure 10.12   Initialization routine for aggregate alias lists.

Interprocedural Analysis

alias lists". On an alias list, a global is always an atom, but a formal may be either a pointer to another alias list, or a pointer and an atom. In the dual case, we prefix that formal by a star.

To initialize alias lists, we temporarily postpone placing anything on an alias list of a global, and call INITIALIZE in Figure 10.12. Since aggregate aliasing patterns are relative to a particular procedure, we must restrict the transitive closure accordingly. If globals G1 and G2 are aliases within procedure S, then they are not necessarily aliases in all procedures. That is, we want to avoid placing G2 on G1's list, and G1 on G2's list.

Example 10.10 Consider the SIMPL-T program in Figure 10.13(i). Here is the action for the **CALL** statements.

CALL Statement	Action
$P(A, B, C)$	Place A on X's list, B on Y's list, and C on Z's list.
$P(D, I, I)$	Place D on X's list, $*Z$ on Y's list, and $*Y$ on Z's list.
$Q(X, X, F)$	Place $*V$ on U's list, and $*U$ on V's list.
$Q(Z, E, E)$	Place Z on U's list, and E on V's list.

To form the explicit transitive closure of those alias lists, the SCC transitive closure routine of the previous section can be employed. Example 10.11 gives an intuitive explanation of this process for the "throw-away" but illustrative example SIMPL-T program in Figure 10.13.

```
INT A, B, C, D, D, E, F
PROC DONOTHING
  INT I
  CALL P(A, B, C)
  CALL P(D, I, I)
  RETURN
PROC P (REF INT X, REF INT Y, REF INT Z)
  CALL Q(X, X, F)
  CALL Q(Z, E, E)
  RETURN
PROC Q (REF INT U, REF INT V, INT W)
  RETURN
START DONOTHING
          (i)
```

$X: A, D$
$Y: B, *Z$
$Z: C, *Y$
$U: X, *V, Z$
$V: X, *U, E$

(ii)

Figure 10.13 An example to illustrate the initialization of aggregate aliasing patterns. (i) A "throw-away" Example of a SIMPL-T Program. (ii) Initial Alias Lists of REF Formals.

Example 10.11 Form a directed graph G called an "aliasing graph", where each formal is represented by a node, and there is an arc from X to Y iff Y appears (either starred or unstarred) on X's list. Find the strongly connected components (SCCs) of G and at the same time form the set union of all globals of each SCC. See Figure 10.14. Form the acyclic condensation of G, G', by shrinking each SCC of G into a node. Now, if there is a path from node i to node j in G', union the global set for j to that of i. Finally, the alias list for each formal is the global list associated with the SCC containing that formal and those formals that originally appeared starred on its list. For example, the alias lists for U in Figure 10.14 is $\{A, B, C, D, E, V\}$.

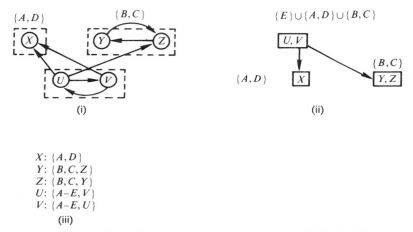

$X: \{A, D\}$
$Y: \{B, C, Z\}$
$Z: \{B, C, Y\}$
$U: \{A-E, V\}$
$V: \{A-E, U\}$
(iii)

Figure 10.14 Transitive closure of aggregate aliasing patterns.
(i) SCCs of aliasing graph; (ii) acyclic condensation of aliasing graph;
(iii) aggregate aliasing lists.

Here is the complete algorithm.

ALGORITHM 10.2 *Computation of aggregate aliasing patterns.*

(1) **call** INITIALIZE
(2) /* Form aliasing graph. */
Form the aliasing graph, and represent it as successor lists. (Each node represents a **REF** formal, and there is an arc (X, Y) iff Y is on X's initial alias list from (1).)
For each node X let $A(X)$ be the set of globals on X's alias list from step (1).
(3) /* Perform transitive closure via depth-first search. */

Apply Algorithm 10.1 to the aliasing graph and the A sets. (Use **setunion** instead of **bvor**.)

(4) The final alias list for formal X is the set union of $A(\text{SCCROOT}(X))$ and those formals on $A(X)$ from step (1) that were starred.

(5) **for** each alias list $A(X)$ of each formal X **do**
   if global G is on $A(X)$ **then** add X to $A(G)$ **endif**
endfor
/* We assume here that the "home" procedure of each formal is known, and thus formals with the same names can be distinguished. Note that in SIMPL-T no global is ever on the alias list of another global. */

(6) Form a list of all **EXT** and **ENTRY** globals. If any alias list contains an **EXT** or **ENTRY** global, give that list a special mark.
/* Rather than burden each list containing at least one **EXT** or **ENTRY** global with all such globals, a special mark is equally informative. */ □

10.5 Intraprocedural Analysis

10.5.1 The Equivalence Relation

To identify redundant computations, expressions, assignments, address calculations, and function calls, it is necessary to keep a list of those expressions and variables that have already been computed and whose computed values are still available (i.e., those that always reach a given point). One very efficient method is to employ an equivalence relation.

It is apparent that merely listing known expressions in the form they are encountered is not, in general, sufficient. Both commutativity of such operators as '+' and '*', and equivalence implied by ':=' must be considered.

An *equivalence relation (ER)* is a set of zero or more disjoint equivalence classes. An *equivalence class (EC)* is a set of one or more items, all of which are known to be equivalent. An item can be a variable, a constant, an expression, an address, or a function call. In our implementation, each EC has a unique "value number" associated with it, and any expression in an EC has its operands replaced by their respective value numbers. Thus an expression need be mentioned only once in an EC, rather than with all possible combinations of equivalent operands. Commutative operators are handled by always storing them with the smaller numbered operand first. Finally, it is not necessary to list expression temporary numbers in an EC since we dynamically renumber the temporaries to correspond to the EC number. Because there can never be more than one expression in an EC, this does not lead to ambiguous numbering (a second expression in an EC would be redundant by definition and hence eliminated).

When a variable or constant is encountered, we make sure that it has an entry in the ER. When an expression is encountered we determine the EC numbers of its operands, put them in increasing order if the operation is commutative, and search the ER. If an equivalent expression is found, then the current one is redundant. We remember that all future references to the result temporary of the current quad must be replaced by the EC number of the expression already in the ER. We must also set the "save" flag (QSFLG) in the last previous quad use of that EC number. If the current expression is unique, we create a new EC and put the modified form of the expression into it. The result temporary in the quad corresponding to the expression is set to the new EC number, and we remember that all future references to that temporary must be changed to that number.

10.5.2 Redundant Computations in Straight-Line Code

To check an assignment statement, we find the EC numbers of the left-hand and right-hand sides. If they are equal, then the assignment is redundant. Otherwise, we remove the left-hand side from its place in the ER and put it in the same EC as the right-hand side. References to possible aliases of the left-hand side must also be removed from the ER.

To process a procedure or function call, we must remove all ER entries that refer to variables that may be modified by the procedure or function. This is accomplished using the appropriate PKILL bit vector. In addition, all references in the ER to call-by-reference parameters are removed. That is, we assume that all call-by-reference parameters are modified.

At the end of each procedure, we must empty out the entire ER. This is necessary because the SIMPL-T compiler allocates temporary storage to each procedure separately.

Removal of an ER entry is a recursive process. Removal of one entry may make an EC empty, in which case all entries of other ECs that refer to this one must also be removed, and so on.

Updating the ER inside other control structures of SIMPL-T is covered in subsequent sections.

Example 10.12 Figure 10.15 shows how an ER is used to detect redundant computations. The code sequence given illustrates the basic ER operations. We follow the convention of using vertical lines to separate ECs and a parenthesized number to refer to an EC. EC $\#i$ is introduced by '(i):'.

Intraprocedural Analysis

Statements	ER and Comments
$A := B + C$	$\{(1):(2)+(3), A \mid (2): B \mid (3): C\}$
	Expression $B + C$ is stored as $(2)+(3)$.
$D := B$	
$E := C$	$\{(1): (2)+(3), A \mid (2): B, D \mid (3): C, E\}$
$A := E + D$	After replacing E, D by their value numbers with the smaller one first, $E + D$ is seen to be redundant. Since the value numbers of A and $E + D$ are 1, the assignment is also redundant.
	$\{(1): (2)+(3), A \mid (2): B, D \mid (3): C, E\}$
$D := E + A$	The operands are switched in $E * A$
	$\{(1): (2)+(3), A \mid (2): B \mid (3): C, E \mid (4): (1)*(3), D\}$
$B := (B + C) * C$	The expression $B + C$ is redundant, and so is the multiplication.
	$\{(1): A \mid (3): C, E \mid (4): (1)*(3), D, B\}$
	Emptying EC #2 requires removal of $(2)+(3)$ from EC 1.

Figure 10.15 Use of the equivalence relation.

A slight modification of the ER concept is necessary to handle arrays. In the SIMPL-T compiler, all temporaries generated by the ARRAYLOC quad are held as addresses until they are used. Thus, they must be kept in separate ECs from all other items.

Example 10.13 Figure 10.16 illustrates how array addressing is handled.

Statements	ER and Comments
$X := A(I)$	$\{(1): I \mid (2): A((1)) \mid (3): X\}$
	X is not in the same class as $A(I)$ since $A(I)$ is an address.
$A(I) := A(J)$	The address $A(I)$ is redundant, but is not removed from the ER even though its value has changed.
	$\{(1): I \mid (2): A((1)) \mid (3): X \mid (4): J \mid (5): A((4))\}$
$I := 1$	$\{(3): X \mid (4): J \mid (5): A((4)) \mid (6): 1, I\}$
	EC #2 is removed since the address is no longer valid. I is put into EC #6 with 1.

Figure 10.16 Array addressing in the equivalence relation.

To detect redundant function calls, we store as expressions all function calls with at most two arguments that do not have side effects. [The existence (or nonexistence) of side effects has already been determined in PASS1.] Although it is possible to accommodate function calls with more than two arguments, we have chosen to restrict the number to two so as to keep our data structure very simple. With at most two arguments, we can treat function calls like binary expressions.

Example 10.14 Figure 10.17 shows the result of a function call of one argument.

Statement	ER
$X := F(Y)$	$\{(1): F((2)), X \mid (2): Y\}$

Figure 10.17 Function call in the equivalence relation.

To implement the ER, we employ a doubly linked list of EC header nodes, each of which points to a doubly linked list of EC items. The double linkage allows easy deletion of items and classes. Since at most one expression exists in any EC, it is stored in the EC header. If a variable or constant is added to an EC, a field in the symbol table entry of that item points to its location in the EC. This allows us to quickly determine the EC number of a given variable or constant.

10.5.3 Redundant Computations in WHILE Loops

To detect redundant computations in a **WHILE** loop in one top-down pass, it is necessary to make certain general assumptions both at loop entry and at loop exit.

On entry to the loop we must remove from the ER all references to variables that are defined anywhere within the loop. Otherwise an expression might be considered redundant on a first iteration path that is not redundant on a loop-iteration path. To implement this, a WKILL bit vector has been constructed for each loop in PASS1. It is now a simple matter in PASS2 to interrogate the proper bit vector on **WHILE** loop entry.

A second problem is encountered on loop exit, since it is possible for a **WHILE** loop to be executed zero times. Any computations performed within the body of the loop may not in fact have been done, so it is necessary to remove all such entries from the ER. To accomplish this, a global variable NESTING is incremented on loop entry and decremented on loop exit. Each entry in the ER has a NESTDEPTH field that is set to the value of NESTING at the time of its creation. At the end of a loop, the ER is sequentially searched and all entries with NESTDEPTH(ENTRY) = NESTING are removed.

Intraprocedural Analysis

In addition to normal loop termination, it is possible to leave a **WHILE** loop by means of the **EXIT** statement, which transfers control to the next statement after the loop. Surprisingly, this does not affect the ER in any way. Code within the loop occurring after the **EXIT** may or may not ever be reached, but it can still be analyzed normally. At the end of the loop we have removed from the ER all entries that were modified anywhere within it, and we are deleting all new entries created by the loop, so the possibility of early termination is covered by these operations.

Example 10.15 Figure 10.18 illustrates the assumptions that must be made in analyzing **WHILE** loops. Neither $B + C$ nor $G(C)$ can be considered redundant.

Statements	ER
$A := B + C$	$\{(1): B\|(2): C\|(3): A, (1)+(2)\}$
WHILE $A > 0$ **DO**	$\{(1): A\|(2):0\|(3): (1)>(2)\}$
$\left.\begin{array}{l}A := B + C \\ C := F(B) \\ B := G(C)\end{array}\right\}$	$\{(1): C\|(2): 0\|(3): G((1)), B\|(6): A\}$
ENDWHILE	$\{(2): 0\}$
$X := G(C)$	$\{(1): C\|(2): 0\|(3): G((1)), X\}$

Figure 10.18 Updating the ER in WHILE loops.

10.5.4 Redundant Computations in IF and CASE Statements

It is possible to process **IF** and **CASE** control structures by simply updating the ER at the end of each clause. Since that clause may not be executed at run-time, all ER entries involving computations in the clause must be removed. This is accomplished by incrementing and decrementing the variable NESTING for **IF** and **CASE** statements as well as **WHILE** statements. At the end of each clause, all ER entries with NESTDEPTH-(ENTRY) = NESTING are removed.

However, this method will miss many redundant computations within succeeding clauses. A preferable procedure is to return the ER to its state prior to the first clause each time we enter a new clause.

Example 10.16 The redundant expression $A + B$ will be missed in Figure 10.19 unless the ER is returned to its state before the **THEN**-clause prior to analyzing the **ELSE**-clause.

To accomplish this we "virtually kill" ER entries that would normally be removed while processing **IF** and **CASE** clauses. But we still must remove

```
IF (A + B) = 0 THEN
    X := 1
    A := 1
ELSE
    X := A + B
ENDIF
```

Figure 10.19 Redundant expression in IF-THEN-ELSE statement.

all newly created entries at the end of a clause as described above. Only entries generated prior to the **IF** or **CASE** statement are left. A global variable INSIDEIFORCASE is incremented on entry to an **IF** or **CASE** statement and is decremented on statement exit. Whenever an EC entry should be removed but INSIDEIFORCASE > 0, we set a field called CLAUSEDEPTH to the value of INSIDEIFORCASE. This marks the entry as being "virtually killed", and it will be ignored in searches for redundant computations. If an entire EC is made empty in this fashion, then all references to it by other entries of the ER must also be virtually killed, and so on, just as in the case of actual removal.

When the next clause in the statement is reached, all entries that were virtually killed in the previous clause (i.e., CLAUSEDEPTH(ENTRY) = INSIDEIFORCASE) have an UNKILLED bit set to 1. These entries are now considered valid again. This action, coupled with the removal of all entries created in the previous clause, recreates the state of the ER at the beginning of the statement.

At the end of the final clause, all ER entries created in that clause must be removed, just as in other clauses, and the variables NESTING and INSIDEIFORCASE are decremented. We must now do something with those ER entries marked "virtually killed" with CLAUSEDEPTH-(ENTRY) = INSIDEIFORCASE + 1. If INSIDEIFORCASE = 0, then the statement is not nested within an outer **IF** or **CASE**, and so we remove these entries from the ER. If INSIDEIFORCASE > 0, then the entries have been killed not only within the statement just exited, but also within a surrounding one. We therefore decrement the CLAUSEDEPTH field of these entries by 1, and zero the UNKILLED bit (if it is on). These entries are now marked as virtually killed in the surrounding clause, which is what we want.

To summarize, the following steps are taken:

(1) At statement entry
    (a) NESTING : = NESTING + 1
    (b) INSIDEIFORCASE : = INSIDEIFORCASE + 1
(2) Within clauses: (i.e., INSIDEIFORCASE > 0)

Intraprocedural Analysis

   (a) Each time an ER entry is invalidated, set
   CLAUSEDEPTH(ENTRY) : = INSIDEIFORCASE
   UNKILLED(ENTRY) : = 0 (in case the entry was
   virtually killed and then unkilled in a
   surrounding statement)
   (b) Newly created ER entries have
   NESTDEPTH(ENTRY) : = NESTING
(3) At clause exit:
   (a) All ER entries with NESTDEPTH(ENTRY) = NESTING are
   unconditionally removed.
   (b) All ER entries with CLAUSEDEPTH(ENTRY) = INSIDEIFORCASE have UNKILLED(ENTRY) : = 1 (not
   done on exit from final clause in statement)
(4) At statement exit:
   (a) All ER entries with NESTDEPTH(ENTRY) = NESTING are
   unconditionally removed.
   (b) INSIDEIFORCASE : = INSIDEIFORCASE − 1
   (c) **if** INSIDEIFORCASE = 0 **then**
   All entries with CLAUSEDEPTH(ENTRY) = 1 are removed.
   else
   All entries with CLAUSEDEPTH(ENTRY) = (INSIDEIFORCASE + 1) have
   CLAUSEDEPTH(ENTRY) : = INSIDEIFORCASE
   UNKILLED(ENTRY) : = 0
   endif
   (d) NESTING : = NESTING − 1

A slightly more efficient algorithm is obtained by doing step 4(b) just before an **ELSE**-clause if one exists. Since the **ELSE**-clause is the last one in the statement, we do not need to virtually kill ER entries within that clause unless it is nested within a surrounding statement. Decrementing INSIDEIFORCASE before the **ELSE**-clause allows us to skip step 2(a) if INSIDEIFORCASE = 0. We can further improve the algorithm by noting in PASS1 which **IF** statements have an **ELSE**-clause. Those **IF** statements that have no **ELSE** clause do not require us to save the ER environment, so steps 1(b), 3(b), 4(b), and 4(c) can be omitted. [Step 2(a) may still be executed if we are within a surrounding clause.]

10.5.5 Invariant Code in WHILE Loops

Removal of invariant code from loops can be performed in parallel with redundant quad removal. We can determine the movability of an expression quad by checking its operands against the WKILL bit vector and the equivalence relation. Expressions with variable operands are movable only if those variables are not changed anywhere within the loop. This information is easily obtained by checking the proper bit in the WKILL vector. Expressions with temporary operands are movable only if those temporaries are generated outside of the **WHILE** loop. This is determined by checking the ER entry corresponding to each temporary operand. If NESTDEPTH(ENTRY) < NESTING, then the temporary was generated before the loop and hence the expression using it may be movable.

When we move an expression out of a loop, we must decrement its NESTDEPTH field by one to indicate that it is generated outside of the **WHILE** loop. This allows us to move other expressions that depend on it. Note that we cannot move assignment quads since the loop may not be executed. As in redundant expression analysis, **EXIT** statements do not affect the results.

When expressions are moved up past other code, their result temporaries must be renumbered so that they do not conflict.

Example 10.17 In Figure 10.20, the quad $(ARRAYLOC, A, I, t1)$ can be moved outside the loop, but its result temporary must be renumbered in order not to conflict with $(+, X, 1, t1)$.

Statements	Quads
WHILE 1 **DO**	(WHILE,,,)
$X := X + 1$	$(+, X, 1, t1)$
	$(:=, t1,, X)$
$Y := A(I)$	$(ARRAYLOC, A, I, t1)$
	$(:=, *t1,, Y)$
$\vdots$	$\vdots$

Figure 10.20 Movable code in a WHILE loop.

This problem is further complicated by the fact that we want to recursively move code out of nested loops without backtracking to again renumber temporaries. For example, suppose we have reached the end of an inner **WHILE** loop, and we now see that some of its invariant code is movable with respect to an outer loop. It might be the case that we would have to re-renumber result temporaries in that code, and then go back into the inner loop to change corresponding uses of those temporaries. To avoid this, we can keep track of the largest temporary number ever used so far within one or more nested loops. Whenever an expression is initially moved, its result temporary is set to the next available number so that it

Discussion

cannot possibly conflict. Thus we do not have to backtrack when renumbering temporaries. (Incidently, we compact the range of numbers used for temporaries as we output the final quads by (again) renumbering them using a heap[4] to implement a priority queue that contains available temporary numbers.)

The actual moving of expressions can be done in either of two ways. One way is to delete movable code from the quad file and put it in a list along with information about where it goes. Another pass through the file is then required to reinsert the code. A second method is to hold quads in a linked output buffer large enough to include entire loops. Quadruples can then be moved by redirecting links in the buffer.

10.6 Discussion

In summary, we have sketched some ideas on the design of a modest code improver for a structured programming language. Our improver operates on an intermediate-level program representation (quadruples) that still reflects the high-level control flow structures of the language. A slightly modified recursive descent approach is used. First interprocedural information is collected and processed, then intraprocedural analysis is performed. The basic data structures constructed and manipulated for the interprocedural part are the call graph, bit vectors indicating the variables whose values may be modified by each procedure and loop, and aggregate alias lists. For the intraprocedural part, the basic data structure is a doubly linked list of doubly linked lists, for the equivalence relation.

Preliminary experiments indicate between 5% and 15% reduction in the number of quads. (Extensive execution time experiments have not been run yet.) We feel that this performance is good considering the very simple design of the improver. However, rather than expend more energy making an intermediate-level improver more sophisticated, we feel that there is a point of diminishing return on intermediate-level improvement. For thorough code improvement, the low-level code should be scrutinized too.

* * *

There are (as indicated in Section 1.2 of this book) other applications of flow analysis besides code improvement. It is this variety of applications that indicates to us that flow analysis techniques will be more pervasively used in future programming language processors.

[4]See Aho, Hopcroft, and Ullman [1975], for example.

Exercises

10.1 PASCAL has **goto** statements and strongly typed pointer variables. Assuming these features occur infrequently, can the approach taken in this chapter for SIMPL-T be adapted (with slight modification) to PASCAL?

10.2 Is there an aliasing problem with call-by-value-and-result or call-by-name?

10.3 Is there a simple way of incorporating in the design described in this chapter the hoisting and sinking of computations from **IF-THEN-ELSE** statements?

Bibliographic Notes

The study described in this chapter is from Hecht and Shaffer [1976]. This work is in part a follow-up of that done by Zelkowitz and Bail [1974]. Geshke [1972], and Wulf et al. [1975] describe a recursive descent code improver for the structured programming language BLISS.

Loveman [1976] is an excellent source for source-to-source program transformations. Basili and Turner [1975] give a complete description of SIMPL-T.

Some outstanding articles on interprocedural analysis and the problem of dynamic aliasing include Spillman [1971], Allen [1974, 1975b], Rosen [1975a, 1975b, 1976], Lomet [1975], and Barth [1977]. Rosen [1975b] gives a theoretical foundation for high-level flow analysis (i.e., sans flow graphs, and essentially top-down).

Algorithm 10.1 comes from Tarjan [1972].

The idea of using an equivalence relation for flow analysis is due to Kildall [1973]. Value numbers are described in Cocke and Schwartz [1970].

Bibliography

Abbreviations Used
ACM   *A*ssociation for *C*omputing *M*achinery
CACM   *C*ommunications of the *ACM*
IFIP   *I*nternational *F*ederation for *I*nformation *P*rocessing
JACM   *J*ournal of the *ACM*
JCSS   *J*ournal of *C*omputer and *S*ystem *S*ciences
SIAM   *S*ociety for *I*ndustrial and *A*pplied *M*athematics
STOC   *A*CM *S*ymposium on *T*heory *O*f *C*omputing
POPL   *P*roceedings of ACM Symposium on *P*rinciples *O*f *P*rogramming *L*anguages

AHO, A.V., HOPCROFT, J. E., and ULLMAN, J.D. [1974]. *The Design and Analysis of Computer Algorithms*, Addison-Wesley, Reading, Mass.

AHO, A.V., SETHI, R., and ULLMAN, J.D. [1972]. "Code Optimization and Finite Church-Rosser Systems", *Design and Optimization of Compilers*, R. Rustin, Ed., Prentice-Hall, Englewood Cliffs, N. J., 89–106.

AHO, A.V., and ULLMAN, J.D. [1972]. *The Theory of Parsing, Translation, and Compiling*, Volume 1: *Parsing*, Prentice-Hall, Englewood Cliffs, N. J.

AHO, A.V., and ULLMAN, J.D. [1973]. *The Theory of Parsing, Translation, and Compiling*, Volume 2: *Compiling*, Prentice-Hall, Englewood Cliffs, N. J.

AHO, A.V., and ULLMAN, J.D. [1975]. "Node Listings for Reducible Flow Graphs", *7th STOC*, 177–185.

AHO, A.V., and ULLMAN, J.D. [1977]. *Principles of Compiler Design*, preliminary manuscript.

ALLEN, F.E. [1969]. "Program Optimization", *Annual Review of Automatic Programming*, 5, 239–307.

ALLEN, F.E. [1970]. "Control Flow Analysis", *SIGPLAN Notices*, 5:7, 1–19.

ALLEN, F.E. [1971]. "A Basis for Program Optimization", *Proc. IFIP Congress* 71, North Holland Publishing Co., Amsterdam, Holland, 385–390.

ALLEN, F.E. [1974]. "Interprocedural Data Flow Analysis", *Proc. IFIP Congress* 74, North Holland Publishing Co., Amsterdam, Holland, 398–492.

ALLEN, F. E. [1975a]. "Bibliography on Program Optimization", IBM Research Report RC-5767, T.J. Watson Research Center, Yorktown Heights, N. Y.

ALLEN, F.E. [1975b]. "Interprocedural Analysis and the Information Derived by It", *Programming Methodology*, Springer-Verlag, Berlin, Germany, 291–321.

ALLEN, F.E., and COCKE, J. [1972]. "Graph-Theoretic Constructs for Program Control Flow Analysis", IBM Research Report RC-3923, T. J. Watson Research Center, Yorktown Heights, N. Y.

ALLEN, F.E., and COCKE, J. [1976]. "A Program Data Flow Analysis Procedure", *CACM* 1:3, 137–147.

ASHCROFT, E., and MANNA, Z. [1971]. "The Translation of 'GO TO' Programs to 'WHILE' Programs", *Proc. IFIP Conf.* 71, North Holland Publishing Co., Amsterdam, Holland.

BACKHOUSE, R.C. [1976]. "Comparisons of Iterative Techniques and of Elimination Techniques in Closure Algorithms", Technical Report 2, Dept. of Computer Science, Heriot-Watt Univ., Edinburgh, Scotland.

BARTH, J.M. [1977]. "An Interprocedural Data Flow Analysis Algorithm", *4th POPL*, Los Angeles, Calif., 119–131.

BASILI, V.R., and TURNER, A.J. [1975]. "SIMPL-T: A Structured Programming Language", Computer Note CN-14.1, Univ. of Maryland, College Park, Md.

BAUER, A.M., and SAAL, H.J. [1974]. "Does APL Really Need Run-Time Checking?", *Software Practice and Experience*, **4**, 129–138.

BIRKHOFF, G. [1967]. *Lattice Theory*, American Mathematical Society Colloquium Publications, Vol. XXV, Providence, R. I.

BITTINGER, M.L. [1972]. *Logic and Proof*, Addison-Wesley, Reading, Mass.

BOHM, C., and JACOPINI, G. [1966]. "Flow Diagrams, Turing Machines and Languages with Only Two Formation Rules", *CACM*, **9**:5, 366–371.

COCKE, J. [1970]. "Global Common Subexpression Elimination", *SIGPLAN Notices*, **5**:7, 20–24.

COCKE, J. [1971]. "On Certain Graph-Theoretic Properties of Programs", IBM Research Report RC-3391, T. J. Watson Research Center, Yorktown Heights, N. Y.

COCKE, J., and MILLER, R. E. [1969]. "Some Analysis Techniques for Optimization of Computer Programs", *Proc. 2nd Hawaii Conf. on System Sciences*, 143–146.

COCKE, J., and SCHWARTZ, J. T. [1970]. *Programming Languages and Their Compilers*, Courant Institute of Mathematical Sciences, New York, N.Y.

COOPER, D.C. [1968]. "Some Transformations and Standard Forms of Graphs, with Applications to Computer Programs", *Machine Intelligence* 2, American Elsevier, New York, 21–32.

COOPER, D.C. [1971]. "Programs for Mechanical Program Verification", *Machine Intelligence* 6, American Elsevier, New York, 43–62.

DIJKSTRA, E.W. [1968]. "Goto Statement Considered Harmful", *CACM*, **11**:3, 147–148.

EARNEST, C.P., BALKE, K.G., and ANDERSON, J. [1972]. "Analysis of Graphs by Ordering of Nodes", *JACM*, **19**:1, 23–42.

ENGELER, E. [1971a]. "Structure and Meaning of Elementary Programs", *Symposium on Semantics of Algorithmic Languages*, Springer-Verlag, New York, N.Y., 89–101.

ENGELER, E. [1971b]. "Algorithmic Approximations", *JCSS*, **5**, 61–82.

FARROW, R., KENNEDY, K., and ZUCCONI, L. [1976]. "Graph Grammars and Global Program Data Flow Analysis", *17th Annual Symposium on Foundations of Computer Science*, Houston, Texas.

FONG, A.C. [1977]. "Generalized Common Subexpressions in Very High Level Languages", *4th POPL*, Los Angeles, Calif., 48–57.

FONG, A.C., KAM, J.B., and ULLMAN, J. D. [1975]. "Applications of Lattice Algebra to Loop Optimization", *2nd POPL*, Palo Alto, Calif., 1–9.

FONG, A.C., and ULLMAN, J.D. [1976]. "Finding the Depth of a Flow Graph", *8th STOC*, Hershey, Pennsylvania, 121–125.

FOSDICK, L.D., and OSTERWEIL, L.J. [1976]. "Data Flow Analysis in Software

Reliability", *ACM Computing Surveys*, **8**:3, 305–330.

FREDERICKSON, G.N. [1975]. "Refinements to Aho and Ullman's Node Listing Algorithm", Technical Report No. 404, Dept. of Computer Science, Univ. of Maryland, College Park, Md.

GANNON, J.D., and HECHT, M.S. [1977]. "An $O(n^3)$ Algorithm for Parsing a Proper Program into Its Prime Subprograms", Dept. of Computer Science, Univ. of Maryland, College Park, Md.

GESCHKE, C.M. [1972]. *Global Program Optimization*, Ph.D. Dissertation, Dept. of Computer Science, Carnegie-Mellon Univ., Pittsburgh, Pa.

GRAHAM, S.L. and WEGMAN, M. [1976]. "A Fast and Usually Linear Algorithm for Global Flow Analysis", *JACM*, **23**:1, 172–202.

GRATZER, G. [1971]. *Lattice Theory: First Concepts and Distributive Lattices*, W. H. Freeman and Co., San Francisco, Calif.

GRIES, D. [1971]. *Compiler Construction for Digital Computers*, John Wiley and Sons, New York, N. Y.

HARRISON, W. [1975]. "Compiler Analysis of the Value Ranges of Variables", IBM Research Report RC-5544, T. J. Watson Research Center, Yorktown Heights, N. Y.

HECHT, M.S. [1974]. "Topological Sorting and Flow Graphs", *Proc. IFIP Congress 74*, 494–499.

HECHT, M.S., and SHAFFER, J.B. [1976]. "A Modest Quad Improver for SIMPL-T", Dept. of Computer Science, Univ. of Maryland, College Park, Md.

HECHT, M.S., and ULLMAN, J.D. [1972]. "Flow Graph Reducibility", *SIAM J. Computing*, **1**:2, 188–202.

HECHT, M.S., and ULLMAN, J.D. [1974]. "Characterizations of Reducible Flow Graphs", *JACM*, **21**:3, 367–375.

HECHT, M.S., and ULLMAN, J.D. [1975]. "A Simple Algorithm for Global Data Flow Analysis Problems", *SIAM J. Computing*, **4**:4, 519–532.

HOPCROFT, J.E., and ULLMAN, J.D. [1969]. *Formal Languages and Their Relation to Automata*, Addison-Wesley, Reading, Mass.

HOPCROFT, J.E., and ULLMAN, J.D. [1972]. "An $n \log n$ Algorithm for Detecting Reducible Graphs", *Proc. 6th Annual Princeton Conf. on Information Sciences and Systems*, 119–122.

JAZAYERI, M. [1975]. "Live Variable Analysis, Attribute Grammars, and Program Optimization", Dept. of Computer Science, Univ. of North Carolina, Chapel Hill, N.C.

JONES, N.D., and MUCHNICK, S.S. [1976]. "Binding Time Optimization in Programming Languages: Some Thoughts Toward the Design of an Ideal Language", 3rd *POPL*, Atlanta, Georgia, 77–94.

KAM, J.B. [1973]. "Flow Graph Reducibility of Certain Subclasses of Flow Charts", Technical Report No. 141, Dept. of Electrical Engineering, Princeton Univ., Princeton, N.J.

KAM, J.B. [1975]. *Monotone Data Flow Analysis Frameworks: A Formal Theory of Global Computer Program Optimization*, Ph.D. Dissertation, Dept. of Electrical Engineering, Princeton Univ., Princeton, N. J.

KAM, J.B., and ULLMAN, J.D. [1976]. "Global Data Flow Analysis and Iterative

Algorithms", *JACM*, **23**:1, 158–171.

KAM, J.B., and ULLMAN, J.D. [1975]. "Monotone Data Flow Analysis Frameworks", Technical Report No. 169, Dept. of Electrical Engineering, Princeton Univ., Princeton, N. J.

KASVANOV, V.N. [1973]. "Some Properties of Fully Reducible Graphs", *Information Processing Letters*, **2**:4, 113–117.

KENNEDY, K. [1971]. "A Global Flow Analysis Algorithm", *International J. Computer Math.*, **3**, 5–15.

KENNEDY, K. [1975]. "Node Listings Applies to Data Flow Analysis", *2nd POPL*, 10–21.

KENNEDY, K. [1976]. "A Comparison of Two Algorithms for Global Data Flow Analysis", *SIAM J. Computing*, **5**:1, 158–180.

KILDALL, G.A. [1972]. *Global Expression Optimization During Compilation*, Ph.D. Dissertation, Dept. of Computer Science, Univ. of Washington, Seattle, Wash.

KILDALL, G.A. [1973]. "A Unified Approach to Global Program Optimization", *1st POPL*, Boston, Mass., 194–206.

KNUTH, D.E. [1971]. "An Empirical Study of FORTRAN Programs", *Software Practice and Experience*, **1**:12, 105–134.

KNUTH, D.E. [1974]. *The Art of Computer Programming*, Vol. I: *Fundamental Algorithms*, 2nd Edition, Addison-Wesley, Reading, Mass.

KNUTH, D.E. [1973]. *The Art of Computer Programming*, Vol. III: *Sorting and Searching*, Addison-Wesley, Reading, Mass.

KNUTH, D.E., and FLOYD, R.W. [1971]. "Notes on Avoiding 'GO TO' Statements", *Information Processing Letters*, **1**:1, 23–31.

KOSARAJU, S.R. [1974]. "Analysis of Structured Programs", *JCSS*, **9**:3, 232–255.

KOU, L.T. [1975]. "On Live-Dead Analysis for Global Data Flow Problems", IBM Research Report, T. J. Watson Research Center, Yorktown Heights, N. Y.

LINGER, R.C., and MILLS, H.D. [1977]. *Structured Programming Theory and Practice*, preliminary notes.

LOMET, D.B. [1975]. "Data Flow Analysis in the Presence of Procedure Calls", RC-5728, IBM T. J. Watson Research Center, Yorktown Heights, N. Y.

LOVEMAN, D.B. [1976]. "Program Improvement by Source to Source Transformation", *3rd POPL*, Atlanta, Georgia, 140–152.

LOWRY, E., and MEDLOCK, C.W. [1969]. "Object Code Optimization", *CACM*, **12**:1, 13–22.

MARKOWSKY, G. [1974]. Unpublished memorandum.

NEWMAN, M.H.A. [1942]. "On Theories with a Combinatorial Definition of 'Equivalence'", *Annuals of Mathematicsics*, **43**:2, 223–243.

OSTERWEIL, L.J., and FOSDICK, L.D. [1974]. "Data Flow Analysis as an Aid in Documentation, Assertion Generation, Validation, and Error Detection", Report CU-CS-055-74, Computer Science Dept., Colorado Univ., Boulder, Colo.

PAVLIDIS, T. [1972]. "Linear and Context Free Graph Grammars", *JACM*, **19**:1, 11–22.

PETERSON, W., KASAMI, T., and TOKURA, N. [1973]. "On the Capabilities of While, Repeat, and Exit Statements", *CACM*, **16**:8, 503–512.

PRATT, T.W. [1975]. *Programming Languages: Design and Implementation*, Prentice-Hall, Englewood Cliffs, N. J.

PREPARATA, F.P., and YEH, R.T. [1974]. *Introduction to Discrete Structures*, Addison-Wesley, Reading, Mass.
PROSSER, R.T. [1959]. "Applications of Boolean Matrices to the Analysis of Flow Diagrams", *Proc. AFIPS* 1959 *Eastern Joint Computer Conf.*, Spartan Books, Washington, D. C., 133–138.
PURDOM, P.W., and MOORE, E.F. [1972]. "Immediate Predominators in a Directed Graph", *CACM*, **15**:8, 777–778.
ROSEN, B.K. [1975a]. "Data Flow Analysis for Recursive PL/I Programs", IBM Research Report RC-5211, T.J. Watson Research Center, Yorktown Heights, N. Y.
ROSEN, B. K. [1975b]. "High Level Data Flow Analysis, Parts 1 and 2", IBM Research Report RC-5598, T. J. Watson Research Center, Yorktown Heights, N. Y.
ROSEN, B.K. [1976]. "Data Flow Analysis for Procedural Languages", revision of Rosen [1975a].
RUTHERFORD, D.E. [1965]. *Introduction to Lattice Theory*, Hafner Publishing Co., New York, N.Y.
SCHAEFER, M. [1973]. *A Mathematical Theory of Global Program Optimization*, Prentice-Hall, Englewood Cliffs, N. J.
SCHWARTZ, J.T. [1975]. "Automatic Data Structure Choice in a Language of Very High Level", *2nd POPL*, Palo Alto, Calif., 36–40.
SCOTT, D. [1970]. "Outline of a Mathematical Theory of Computation", *Proc. 5th Annual Princeton Conf. on Information Sciences and Systems*, Princeton, N. J., 169–176.
SETHI, R. [1974]. "Testing for the Church-Rosser Property", *JACM*, **21**:4, 671–679.
SPILLMAN, T.C. [1971]. "Exposing Side-Effects in a PL/I Optimizing Compiler", *Proc. IFIP Congress* 71, North Holland Publishing Co., Amsterdam, Holland, 376–381.
STRONG, H.R., MAGGIOLO-SCHETTINI, A., and ROSEN, B.A. [1975]. "Recursion Structure Simplification", *SIAM J. Computing*, **4**:3, 307–320.
SZSAZ, G. [1963]. *Introduction to Lattice Theory*, Academic Press, New York, N. Y.
TARJAN, R.E. [1972]. "Depth-First Search and Linear Graph Algorithms", *SIAM J. Computing*, **1**:2, 146–160.
TARJAN, R.E. [1973]. "Finding Dominators in Directed Graphs", *Proc. 7th Annual Princeton Conf. on Information Sciences and Systems*, 414–418.
TARJAN, R.E. [1974]. "Testing Flow Graph Reducibility", *JCSS*, **9**:3, 355–365.
TARJAN, R.E. [1976], "Iterative Algorithms for Global Flow Analysis", Technical Report STAN-CS-76-547, Computer Science Department, Stanford University, Stanford, Calif.
TENENBAUM, A. [1974]. "Type Determination in a Language of Very High Level", Courant Institute of Mathematical Sciences, Report No. NSO-3, Computer Science Dept., New York Univ., New York, N. Y.
TREMBLAY, J.P., and MANOHAR, R.P. [1975]. *Discrete Mathematical Structures with Applications to Computer Science*, McGraw-Hill, New York, N. Y.
ULLMAN, J.D. [1973]. "Fast Algorithms for the Elimination of Common Subexpressions", *Acta Informatica*, **2**:3, 191–213.
ULLMAN, J.D. [1975]. "Data Flow Analysis", Technical Report No. 179, Dept. of Electrical Engineering, Princeton Univ., Princeton, N. J. (also in *Second*

US-Japan Computer Conf.)

VYSSOTSKY, V.A. [1973]. Private communication.

WALTER, K.G. [1976]. "Recursion Analysis for Compiler Optimization", *CACM*, **19**:9, 514–516.

WEGBREIT, B. [1974]. "The Treatment of Data Types in EL1", *CACM*, **17**:5, 251–264.

WEGBREIT, B. [1975]."Property Extraction in Well-Founded Property Sets", *IEEE Transactions on Software Engineering*, **1**:3, 270–285.

WULF, W.A., *et al.* [1975]. *The Design of an Optimizing Compiler*, Elsevier North-Holland, Inc., New York.

ZELKOWITZ, M.V., and BAIL, W.G. [1974]. "Optimization of Structured Programs", *Software Practice and Experience*, **4**:1, 51–57.

INDEX

Absorption, 44
Accessible, 35
Active. *See* Live.
Acyclic, 35
Acyclic condensation, 35, 21
AE. *See* Available expressions (AE).
AEBOT, 130, 131
AETOP, 130, 131
AHO, A.V., 14, 26, 27, 37, 49, 71, 104, 113, 114, 116, 123, 127, 159, 215
Algebraic system, 44
ALGOL 60, 3, 7, 13, 25, 70
ALGOL 68, 3
Algorithms, 36–37
    AE$ROUND$ROBIN$VERSION, 140
    depth-first search (DFS), 68, 199–200
    DFS, 68, 200
    dominator, 109–110, 179–180
    Euclid's, 10–14
    FIND$DOMINATORS, 109
    FIND$INTERVALS, 61, 63
    GCD, 11–14
    Graham-Wegman, 24, 53, 72, 120, 127, 128, 141, 147, 185
    INITIALIZE, 204
    interprocedural analysis, 23, 195–207
    interval analysis, 57–67, 146–159, 53, 72, 120, 127
    INTERVAL$ANALYSIS, 154
    intraprocedural analysis, 23, 207–215
    ITERATE, 47
    iterative, 47, 135ff, 173–179
    K1, 176
    K2, 177
    LV$NODE$LISTING$VERSION, 144
    LV$ROUND$ROBIN$VERSION, 140
    PHASE1, 153, 156
    PHASE2, 154, 157
    PHASE3, 157
    topological sorting, 42
    TOPSORT, 42
    ULLMAN's 24, 53, 72, 120, 127, 128, 146, 147, 185
    WORKLIST$VERSION1, 136
    WORKLIST$VERSION2, 137
    WORKLIST$VERSION3, 139
Aliases, 15
Aliasing, 201–207
Alive. *See* Live.
ALLEN, F. E., 23, 26, 27, 71, 72, 104, 113, 114, 123, 145, 146, 159, 181, 217
Ancestor, 36, 68
ANDERSON, J., 113
Antisymmetry, 33
APL, 3, 7
Arc, 35
    back, 68, 93

Index

Arc, 35 (cont'd)
  backward, 86, 93
  cross, 68
  forward, 68, 86
  labeling, 35
  latching, 58
ASHCROFT, E., 123
Associativity, 30, 44
Asymmetry, 33
AVAIL. *See* Available expressions (AE).
Available expressions (AE), 129–131, 135ff, 164ff

BACKHOUSE, R.C., 23, 181
BAIL, W.G., 23, 217
BALKE, K.G., 113
BARTH, J.M., 27
BASIC, 3
Basic block, 13
BASILI, V.R., 38, 49, 217
BAUER, A.M., 7, 26
Binding, 7
Binding time, 7
BIRKHOFF, G., 49
BITTINGER, M.L., 49
Bit vectors, 37
BLISS, 3, 5, 217
Block. *See* Basic block.
BOHM, C., 104
Boolean algebra, 45, 166
Bounded, 45

Call graph, 4, 12, 195–196
  root of, 196
Chain, 41
  length, 41
Child, 36, 68

COBOL, 3
COCKE, J., 14, 23, 26, 27, 71, 104, 113, 114, 123, 145, 146, 159, 181, 217
Collapsibility, 73–78
Commutativity, 30, 44
Compiler, 3
Compile-time, 3
Completion, 34, 75
CONST. *See* Constant propagation.
Constant propagation, 162, 164–165, 167–168
Consumes, 73
Continuous, 47–48
Control flow analysis, 4, 9–13
Control flow graph. *See* Flow graph.
COOPER, D.C., 104, 123
Cycle, 35
  entry node of, 99
  simple, 35
  single-entry, 99
Cycle-free. *See* Acyclic.

DAG. *See* Graph, directed acyclic (DAG).
  of a flow graph, 91–95
Data flow analysis, 4, 13–22
  interprocedural, 4, 19–22, 195–207
  intraprocedural, 4, 15–19, 207–215
Data structures, 37–38
DB (locally exposed definitions), 148
Dead, 17
Decval (decimal value), 173

Index

Definition, 16, 30
  locally exposed (generated), 16
Definition-clear, 16
Definition-use chaining, 19
DeMorgan's law, 45, 134
Depth-first search (DFS), 68, 199–200
Depth-first spanning tree (DFST), 67–70, 198–200
Derived sequence, 64
  length, 95
Descendant, 36, 68
DFST. *See* Depth-first spanning tree (DFST).
DIJKSTRA, E.W., 104
Distance, 88
Distributivity, 45, 47–48, 165–166
DOM, 55, 109–110, 179–180
Dominance, 55–57, 94
  algorithm, 109–110, 179–180
Dominates, 55
  directly, 55
  immediately, 55
  properly, 55
Duality, 45
Dynamic, 3

EARNEST, C.P., 113
EL1, 8
ENGELER, E., 104
Equivalence relation, 34, 203, 207–215
Execution-time. *See* Run-time.
Expressions, 129
  available, *See* Available expressions
  very busy, *See* Very busy expressions

FARROW, R., 23–104
FCR. *See* Finite Church-Rosser (FCR).
Finite Church-Rosser, 74–77, 103
Finite length, 45
Fixed point, 34, 46–47
  maximum (MFP), 173
Flow analysis, 4
Flow graph, 4, 54
  derived, 64
  irreducible, 64
  limit, 64
  reducible (RFG), 64
  reverse, 134
  single-exit, 134
  sparse, 54
  trivial, 64
FLOYD, R.W., 123
Folding. *See* Constant propagation
FONG, A.C., 26, 71, 72, 160ff, 181
FORTRAN, 3, 5, 8, 15, 24, 26, 103, 159, 185
FOSDICK, L.D., 26
Framework
  bit vector, 128–134
  distributive, 165–166
  fast, 168
  instance of a, 165
  monotone, 163–169
FREDERICKSON, G.N., 113

GANNON, J.D., 104
GEN, 129, 131, 164
GESCHKE, C.M., 217
GKUW property, 47–48, 165
Global, 14
GRAHAM, S.L., 5, 23, 47, 49, 104, 145, 161, 168, 181

Graph
  control flow, *See* Flow graph
  directed, 34
  directed acyclic (DAG), 35, 20
  flow, *See* Flow graph
  grammar, 101–102
GRATZER, G., 49
GRIES, D., 26

HARRISON, W., 181
Hash tables, 37
Header, 58, 82
HECHT, M.S., 23, 104, 113, 145, 217
HOPCROFT, J.E., 37, 49, 71, 113, 114, 171, 215

Idempotency, 44
In-degree, 35
Interpreter, 3
Interval, 58–60
  analysis, *See* Algorithms
  order, 58, 65–67, 106–107
  partial, 77
Irreducible (nonreducible), 64, 114–117, 155, 158
Irreflexivity, 33
Iterative algorithm
  general, 173–179
  Kam-Ullman variant, 178
  Kildall's version, 138–139
  Node listing version, 143–144, 176–177
  Round-robin version, 138–142, 176–177
  Worklist version, 135–138, 175–176

JACOPINI, G., 104
JAZAYERI, M., 23
Join, 43
JONES, N.D., 8, 26

KAM, J.B., 26, 45, 47, 49, 72, 104, 160ff, 181
KASAMI, T., 104, 123
KASVANOV, V.N., 104, 113
KENNEDY, K., 23, 71, 104, 112, 113, 127, 145, 159
k-fold product, 34, 75
KILDALL, G.A., 23, 45, 49, 138–139, 145, 160ff, 181, 217
Kill, 16
KILL, 164
KNUTH, D.E., 42, 49, 72, 104, 122, 123, 159
KOSARAJU, S.R., 72, 104, 123
KOU, L.T., 145

Latch. *See* Arc, latching.
Lattice, 43–48
  theory, 40–48
LINGER, R.C., 104
LISP, 3, 7, 25
List
  doubly linked, 37
  singly linked, 37
Live, 17, 19
  definitions, 19, 148
  variables (LV), 16–18, 132–133, 139ff, 148
Local, 14
LOMET, D.B., 27, 217
Loop, 57–58, 100
Loop-connectedness, 70, 95, 97–99, 141–142

Index

LOVEMAN, D.B., 217
LOWRY, E., 26, 71, 145
LV. *See* Live variables (LV).
LVBOT, 16, 18, 19, 132, 133, 139ff
LVTOP, 16, 18, 19, 133, 148ff

MAGGIOLO-SCHETTINI, 26
MANNA, Z., 123
MANOHAR, R.P., 49
MARKOWSKY, G., 113
MEDLOCK, C.W., 26, 71, 145
MEET, 43
   over all paths (MOP), 169–173
Meet-endomorphism, 47–48
MFP. *See* Fixed point, maximum (MFP).
MILLER, R.E., 123
MILLS, H.D., 104
Modified Post's correspondence problem, 170–173
Monotone operation space, 165
Monotonicity, 46
MOORE, E.F., 113
MOP. *See* Meet over all paths (MOP).
MPCP. *See* Modified Post's correspondence problem (MPCP).
MUCHNICK, S.S., 8, 26

NEWMAN, M.H.A., 104
Node, 34
   entry, 100
   exit (terminal), 13, 18, 54, 134, 147
   initial, 13, 54
   string of a, 106
Node listing, 110–112

   minimal, 111
   strong, 111
   weak, 112
Node order, 105
   reasonable, 106
Node splitting, 114–121
NOTDEFINED, 17, 18, 133
NOTKILL, 129ff
NP-complete, 114, 122

O (big-Oh or order-of-magnitude notation), 36
Object language, 3
OSTERWEIL, L.J., 26
Out-degree, 35
OUTSIDE, 16
OUTSIDEDEFS, 152
OUTSIDEUSES, 155

Parent, 36, 68
Partial order, 40, 55
PASCAL 3, 25, 216
Path, 35
   basic, 112
   cycle-free, 35
   simple, 35
PAVLIDIS, T., 104
PB (preserved definitions), 148
PETERSON, W., 104, 123
Pidgin SIMPL, 38–39, 10
Pigeon, 38
PL/I, 3, 5, 24, 25, 70, 185, 187
Poset (*Also see* Partial order), 40
   comparable elements, 40
   maximal element, 40
   minimal element, 40
   one, 40
   zero, 40

POSTORDER, 68, 107–108
Power set algebra, 44
PRATT, T.W., 3, 26
PRED, 35
Predecessor, 35
  lists, 38
Pre-execution, 3
PREPARATA, F.P., 49
Preserved, 16, 78
PRESERVED, 16–18, 132
Procedure, 4
  external, 12, 195
  internal, 12, 195
  start, 4, 12
Property
  absorption, 44
  antisymmetry, 33
  associativity, 30, 44
  asymmetry, 33
  boundedness, 45
  collapsibility, 74
  commutativity, 30, 44
  continuity, 47
  distributivity, 45, 47, 165–170
  fast, 168
  finite Church-Rosser, 74–75
  finite length, 41, 45
  GKUW, 47
  idempotency, 44
  irreducible, 64
  irreflexivity, 33
  meet-endomorphism, 47
  monotonicity, 46, 165
  reasonable, 106
  reducible, 64
  reflexivity, 33
  single-entry, 99–101
  single-exit, 134
  symmetry, 33
  transitivity, 33
  undecidability, 37
  X, 117
PROSSER, R.T., 71
PURDOM, P.W., 113

Quadruples, 10–12, 33, 191–194

RAM. *See* Random access machine.
Random access machine, 36–37, 39
RD. *See* Reaching definitions.
RDBOT, 16, 17, 132
RDTOP, 16, 17, 19, 132, 148ff
Reach, 16
Reaching definitions, 16, 17, 132, 148
Reducibility, 64, 72–104
  by intervals, 57–67
Reflexive and transitive closure, 34, 75
Reflexive closure, 33–34, 75
Region, 82
Regular expression, 116, 170, 173
Representation, 82–83
RFG. *See* Flow graph, reducible (RFG).
  parse of a, 85
ROSEN, B.K., 23, 26, 27, 49, 145, 217
rPOSTORDER, 68, 107–110, 139–142, 173, 177
Run-time, 3
RUTHERFORD, D.E., 49

SAAL, H.J., 7, 26
SCC. *See* Strongly connected component (SCC).

Index

SCHAEFER, M., 104, 113, 123, 145, 159
SCHWARTZ, J.T., 14, 26, 27, 71, 123, 159, 160, 181, 217
SCOTT, D., 49
Semilattice, 45, 47–48, 163–164
SETHI, R., 104
SETL, 7, 24
SHAFFER, J.B., 23, 186, 217
Side effect, 191
SIMPL-T, 3, 5, 7, 188–194
SNOBOL 4, 3, 7, 25
Source language, 3
Space complexity, 36
SPILLMAN, T.C., 27, 217
Static, 3
Step
  bit vector, 37
  elementary, 37
STRONG, H.R., 26
Strongly connected component (SCC), 35, 21, 198–200, 206
Strongly connected region, 58
Structured programming, 38–39, 101–102, 122, 185ff
Subflowgraph, 59
Subgraph
  generated, 35
  partial, 35
  (*), 78–82
SUC, 35
Successor, 35
  lists, 38
Symmetry, 33
SZSAZ, G., 49

TARJAN, R.E., 68, 71, 113, 145, 181, 217
TEMPO, 8

TENENBAUM, A., 26, 49, 145, 160, 181
Time complexity, 36
TOKURA, N., 104, 123
Topsort, 41–43, 21, 66–67, 106–108
Transitive closure, 34, 75, 198–200
Transitivity, 33
Translation-time. *See* Compile-time.
Translator, 3
Tree, 35–36
  dominance, 56–57
TREMBLAY, J.P., 49
TURNER, A.J., 38, 49, 217
Type determination, 7, 162–163
T1, 73
T2, 73–74

UB (locally exposed uses), 147–148
ULLMAN, J.D., 14, 23, 26, 27, 37, 45, 47, 49, 71, 72, 104, 113, 114, 116, 123, 127, 145, 159, 160ff, 181, 215
Undecidable, 37, 170–173
Used, 17
Use, locally exposed, 17

Value numbers, 207, 217
VBE. *See* (VBE).
VBEBOT, 133
VBETOP, 133
Very busy expressions (VBE), 132–133
VYSSOTSKY, V.A., 26

WALTER, K.G., 27

WEGBREIT, B., 8 26, 45
WEGMAN, M., 5, 23, 47, 49, 104, 145, 161, 168, 181
Well-founded, 45
WULF, W.A., 23, 217

XDEFS, 16, 17, 19, 132

XEUSES, 132–133
XUSES, 17–19, 132–133

YEH, R.T., 49

ZELKOWITZ, M.V., 23, 217
ZUCCONI, L., 23, 104